AF505805

Transformations of the State

Series Editors: **Achim Hurrelmann**, Carleton University, Canada; **Stephan Leibfried**, University of Bremen,Germany; **Kerstin Martens**, University of Bremen, Germany; **Peter Mayer**, University of Bremen, Germany.

Titles include:

Joan DeBardeleben and Achim Hurrelmann (*editors*)
DEMOCRATIC DILEMMAS OF MULTILEVEL GOVERNANCE
Legitimacy, Representation and Accountability in the European Union

Klaus Dingwerth
THE NEW TRANSNATIONALISM
Transnational Governance and Democratic Legitimacy

Achim Hurrelmann, Steffen Schneider and Jens Steffek (*editors*)
LEGITIMACY IN AN AGE OF GLOBAL POLITICS

Achim Hurrelmann, Stephan Leibfried, Kerstin Martens and Peter Mayer (*editors*)
TRANSFORMING THE GOLDEN-AGE NATION STATE

Kerstin Martens, Alessandra Rusconi and Kathrin Leuze (*editors*)
NEW ARENAS OF EDUCATION GOVERNANCE
The Impact of International Organizations and Markets on Educational Policy Making

Thomas Rixen
THE POLITICAL ECONOMY OF INTERNATIONAL TAX GOVERNANCE

Peter Starke
RADICAL WELFARE STATE RETRENCHMENT
A Comparative Analysis

Jens Steffek, Claudia Kissling, Patrizia Nanz (*editors*)
CIVIL SOCIETY PARTICIPATION IN EUROPEAN AND GLOBAL GOVERNANCE
A Cure for the Democratic Deficit?

Michael J. Warning
TRANSNATIONAL PUBLIC GOVERNANCE
Networks, Law and Legitimacy

Hartmut Wessler, Bernhard Peters, Michael Brüggemann, Katharina Kleinen-von Kőnigslőw, Stefanie Sifft
TRANSNATIONALIZATION OF PUBLIC SPHERES

Hartmut Wessler (*editor*)
PUBLIC DELIBERATION AND PUBLIC CULTURE
The Writings of Bernhard Peters, 1993–2005

Jochen Zimmerman, Jörg R. Werner, Philipp B. Volmer
GLOBAL GOVERNANCE IN ACCOUNTING
Public Power and Private Commitment

Transformations of the State

Series Standing Order ISBN 978–1–4039–8544–6 (hardback)
Series Standing Order ISBN 978–1–4039–8545–3 (paperback)

You can receive future titles in this series as they are published by placing a standing order. Please contact your bookseller or, in case of difficulty, write to us at the address below with your name and address, the title of the series and the ISBNs quoted above.

Customer Services Department, Macmillan Distribution Ltd, Houndmills, Basingstoke, Hampshire RG21 6XS, England

---

This illustration is taken from the original etching in Thomas Hobbes' *Leviathan* of 1651. Palgrave Macmillan and the editors are grateful to Lucila Muñoz-Sanchez and Monika Sniegs for their help in redesigning the original to illustrate what "transformations of the state" might mean. The inscription at the top of the original frontispiece reads *"non est potestas Super Terram quae Comparetur ei"* (Job 41.33): "there is no power on earth which can be compared to him". In the Bible, this refers to the seamonster, Leviathan. (Original Leviathan image reprinted courtesy of the British Library.)

Transnational Public Governance

Networks, Law and Legitimacy

Michael J. Warning
Attorney
Bremen, Germany

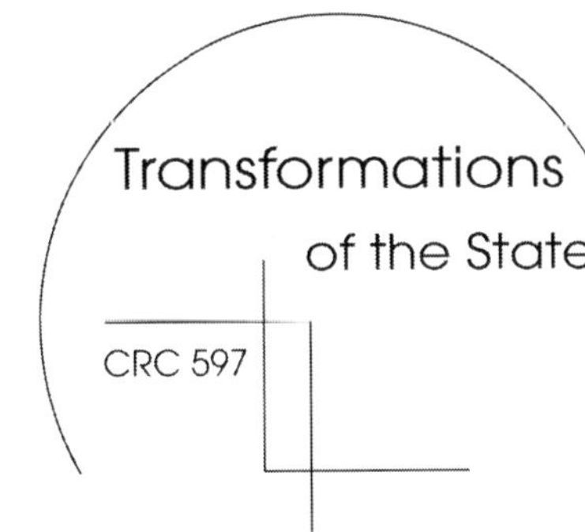

palgrave
macmillan

First published 2009 by
PALGRAVE MACMILLAN

Palgrave Macmillan in the UK is an imprint of Macmillan Publishers Limited, registered in England, company number 785998, of Houndmills, Basingstoke, Hampshire RG21 6XS.

Palgrave Macmillan in the US is a division of St Martin's Press LLC, 175 Fifth Avenue, New York, NY 10010.

Palgrave Macmillan is the global academic imprint of the above companies and has companies and representatives throughout the world.

Palgrave® and Macmillan® are registered trademarks in the United States, the United Kingdom, Europe and other countries.

ISBN-13: 978–0–230–22887–0 hardback

This book is printed on paper suitable for recycling and made from fully managed and sustained forest sources. Logging, pulping and manufacturing processes are expected to conform to the environmental regulations of the country of origin.

A catalogue record for this book is available from the British Library.

A catalog record for this book is available from the Library of Congress.

10 9 8 7 6 5 4 3 2 1
18 17 16 15 14 13 12 11 10 09

Printed in the United States of America

To My Parents

Contents

Part III   The System of Transnational Public Governance

Part IV   The Legitimacy of Transnational Public Governance

Series Editors' Preface

Over the past four centuries, the nation-state has emerged as the world's most effective means of organizing society, but its current status and future are decidedly uncertain. Some scholars predict the total demise of the nation-state as we know it, its powers eroded by a dynamic global economy on the one hand and, on the other, by the transfer of political decisionmaking to supranational bodies. Other analysts point out the remarkable resilience of the state's core institutions and assert that even in the age of global markets and politics, the state remains the ultimate guarantor of security, democracy, welfare, and the rule of law. Do either of these interpretations describe the future of the OECD world's modern, liberal nation-state? Will the state soon be as obsolete and irrelevant as an outdated computer? Should it be scrapped for some new invention, or can it be overhauled and rejuvenated? Or, is the state actually thriving and still fit to serve, just in need of a few minor reforms?

In an attempt to address these questions, the analyses in the *Transformations of the State* series separate the complex tangle of tasks and functions that comprise the state into four manageable dimensions:

- the monopolization of the means of force,
- the rule of law, as prescribed and safeguarded by the constitution,
- the guarantee of democratic self-governance, and
- the provision of welfare and the assurance of social cohesion.

In the OECD world of the 1960s and 1970s, these four dimensions formed a synergetic constellation that emerged as the central, defining characteristic of the modern state. Books in the series report the results of both empirical and theoretical studies of the transformations experienced in each of these dimensions over the past few decades.

Transformations of the State? (Stephan Leibfried and Michael Zürn (eds), Cambridge 2005) and *Transforming the Golden-Age National State* (Achim Hurrelmann, Stephan Leibfried, Kerstin Martens, and Peter Mayer (eds), Basingstoke 2007) define the basic concepts of state transformation employed in all of these studies and provide an overview of the issues addressed. Written by various interdisciplinary teams of political scientists, lawyers, economists, and sociologists, the series tracks the development of the post-World War II OECD state. Here, at last, is a state-of-the-art report on the state of the state and, we hope, a clearer view of its future.

ACHIM HURRELMANN, STEPHAN LEIBFRIED,
KERSTIN MARTENS, AND PETER MAYER
Series Editors

Acknowledgements

This book is a revised and updated version of my dissertation "Transnational Law Through Bureaucracy Networks – The Case of Hazardous Chemicals," which I prepared while working at the "Transformations of the State" Collaborative Research Centre at the University of Bremen.

I would like to express my gratitude to my supervisor, *Prof. Dr. Gerd Winter*. I greatly benefited from his advice, knowledge, and experience and I very much appreciated the opportunity of working with him on various projects. I would also like to thank *Prof. Dr. Josef Falke*, who prepared the second review of the dissertation – an instructive and valuable basis for the improvement of the manuscript. I would like to acknowledge *Prof. Dr. Alfred Rinken*, President of the Constitutional Court of the Free Hanseatic City of Bremen, for sparking my interest in public law during my first semesters at law school. Very special thanks go to *Prof. Dr. Leibfried, Antje Spalink, Dr. Dieter Wolf,* and my colleagues both at the Research Centre for European Environmental Law and the "Transformations of the State" Collaborative Research Centre. I particularly wish to thank *Dr. Martin Herberg*, with whom I conducted the interviews for this book.

I wish to thank *Annette* for her friendship, patience, and encouragement and I am grateful for the support and confidence of my parents, *Elisabeth* and *Winfried*.

Abbreviations

ACC	American Chemistry Council
AGS	*Ausschuß für Gefahrstoffe* (Committee for Hazardous Substances)
APA	Administrative Procedure Act
AU	administrative union
BIAC	Business and Industry Advisory Committee
CAS	Chemical Abstracts Service
CEFIC	*Conseil Européen des Fédérations de l'Industrie Chimique* (European Chemical Industry Council)
CFC	chlorofluorocarbon
CG/HCCS	Coordinating Group for the Harmonization of Chemical Classification Systems
CICAD	Concise International Chemical Assessment Documents
CISG	UN Convention on Contracts for the International Sale of Goods
CoP	Conference of the Parties
CSCE	Conference for Security and Co-operation in Europe
CU	Central Unit
CWM	Chemicals and Waste Management
DDT	Dichloro-Diphenyl-Trichloroethane
DIN	*Deutsches Institut für Normung* (German Institute for Standardization)
DRP	Detailed Review Paper
EC	European Community
ECE	Economic Commission for Europe
ECJ	European Court of Justice
ECOSOC	Economic and Social Council
ED	endocrine disruptor
EDF	Environmental Defense Fund
EDG	Electronic Discussion Group
EHC	Environmental Health Criteria
EHS	Environmental Health and Safety
EPOC	Environmental Policy Committee
EU	European Union
FAO	Food and Agriculture Organization
FRB	Final Review Board

FSC	Forum Standing Committee
GATT	General Agreement on Tariffs and Trade
GHS	Globally Harmonized System of Classification and Labelling of Chemicals
GLP	Good Laboratory Practice
HCB	hexachlorobenzene
γ-HCH	hexachlorocyclohexane, Lindane
HEDSET	Harmonized Electronic Data Set
HPV	High Production Volume
HSG	Health and Safety Guides
IAIS	International Association of Insurance Supervisors
ICC	Intersecretariat Coordinating Committee
ICCA	International Council of Chemical Associations
ICCM	International Conference on Chemicals Management
ICCS	International Conference on Chemical Safety
ICEM	International Federation of Chemical, Energy, Mine and General Workers' Unions
ICJ	International Court of Justice
IFCS	Intergovernmental Forum on Chemical Safety
IFRO	International Financial Regulatory Organizations
ILO	International Labour Organization
IMO	International Maritime Organization
INC	Intergovernmental Negotiating Committee
IO	international organization
IOCC	Inter-Organization Co-ordinating Committee
IOMC	Inter-Organization Programme for the Sound Management of Chemicals
IOSCO	International Organization of Securities Commissions
IPCS	International Programme on Chemical Safety
IRPTC	International Register of Potentially Toxic Chemicals
ISO	International Organization for Standardization
KTA	*Kerntechnischer Ausschuß* (Nuclear Safety Standards Commission)
MAD	Mutual Acceptance of Data
MoU	memorandum of understanding
MEPC	Marine Environment Protection Committee
OECD	Organisation for Economic Co-operation and Development
OEEC	Organisation for European Economic Co-operation
OSCE	Organization for Security and Co-operation in Europe
PAC	Programme Advisory Committee
PCB	polychlorinated biphenyl

PCP	pentachlorophenol
PIC	Prior Informed Consent
POP	persistent organic pollutants
RoP	Rules of Procedure
SAICM	Strategic Approach to International Chemicals Management
SDS	Safety Data Sheet
SIAM	SIDS Initial Assessment Meetings
SIAP	SIDS Initial Assessment Profile
SIAR	SIDS Initial Assessment Reports
SIDS	Screening Information Data Sets
SOP	Standard Operating Procedures
TBT	Technical Barriers of Trade
TCG	Technical Coordinating Group
TNE	Transnational Enterprise
ToR	Terms of Reference
TSCA	Toxic Substances Control Act
TUAC	Trade Union Advisory Committee
UGB-KomE	*Umweltgesetzbuch – Entwurf der unabhängigen Sachverständigenkommission* (Environmental Code – Draft by the Independent Experts Commission)
UN	United Nations
UNCED	United Nations Conference on Environment and Development
UNCETDG	United Nations Committee of Experts on the Transport of Dangerous Goods
UNCETDG/GHS	UN Committee of Experts on the Transport of Dangerous Goods and on the Globally Harmonized System of Classification and Labelling of Chemicals
UNCHE	United Nations Conference on the Human Environment
UNECE	United Nations Commission for Europe
UNEP	United Nations Environment Programme
UNGA	United Nations General Assembly
UNIDO	United Nations Industrial Development Organization
UNITAR	United Nations Institute for Training and Research
UNRTDG	United Nations Recommendations on the Transport of Dangerous Goods
UNSCEGHS	United Nations Economic and Social Council Sub-Committee of Experts on the Globally Harmonized System of Classification and Labelling of Chemicals

UNSCETDG	United Nations/Economic and Social Council Sub-Committee of Experts on the Transport of Dangerous Goods
VCLT	Vienna Convention on the Law of Treaties
WHA	World Health Assembly
WHO	World Health Organization
WNT	Working Group of National Coordinators of Test Guidelines Programme
WSSD	World Summit on Sustainable Development
WTO	World Trade Organization
WWF	World Wide Fund for Nature

1
Introduction

Globalization is currently considered to be one of the most important challenges to the state and international systems, driving structural changes in both. Yet globalization is not entirely new; scholars have distinguished various phases of it, beginning with the rise and interaction of the first empires (Held et al. 2003: 415–35). Furthermore, the term "globalization" lacks a precise definition, despite numerous academic publications shedding light on the phenomenon, although obviously there are a number of processes that mark the contemporary period of globalization, beginning after World War II and accelerating following the fall of communism. Such processes affect the whole planet and almost every aspect of life.

Economic interactions between states have increased and intensified as previously secluded markets have become more and more accessible over the last two decades. Globalization is generally a positive experience for those living in the northern hemisphere who have advantages such as spending their vacations in remote countries, eating at the local McDonald's there if they get homesick or cannot take the local food, and buying exotic produce at their local supermarket after returning home in order to relive their foreign experiences. In general, globalization is marked by high mobility – of individuals, workforce, capital, goods, information, and ideas. However, this mobility has severe repercussions that pose grave dangers to human life and the entire planet. Some of these problems were concealed by the Cold War, while others have emerged only recently.

Highly interwoven markets and the mobility of goods, services, and workforces characterize economic globalization. One benefit is the worldwide availability of virtually everything, everywhere. However, globalization has also resulted in complications, for example, by aggravating the asymmetries in economic wealth between member countries of the Organisation for Economic Co-Operation and Development (OECD) and the rest of the world. Events like the 1997 East Asian financial crisis or the 2008 global financial crisis triggered by the subprime mortgage crisis demonstrate the instability of the world economic system and the potential for future economic disaster. The proliferation of weapons of mass destruction and the launching of the asymmetrical war on terror are by-products of military globalization and reveal the vulnerability of individual states in the current post-Cold War era. The high degree of mobility also impacts upon sanitation, since pathogens, such as the Human Immunodeficiency Virus and the Severe Acute Respiratory Syndrome virus, can now travel the world as quickly and easily as any airplane passenger. Finally, many global environmental problems have emerged in the wake of globalization. Environmental pollutants, such as chlorofluorocarbons (CFCs), disperse around the globe via ocean or air currents, with grave consequences for the world environment.[1]

Such examples give rise to the question of how the problematic aspects of globalization can be overcome. This issue concerns, in particular, the role of the state. One of the key objectives in governing a complex, modern society is the guarantee of both internal and external security, that is, the prevention of harmful events through stabilizing or preserving measures like peacemaking or environmental protection. In the pursuit of such objectives, the state holds the supreme authority and acts as the central institution, determining, issuing, and enforcing regulations from the top down, in a strict hierarchical order (Zürn 1998: 41; 169). Carrying out this type of governance, *governance by government*, is increasingly difficult for individual states as the example of complex, global environmental problems shows. Even 40 or 50 years ago, a polluter's effects were usually limited to his own immediate vicinity; transboundary pollution generally negatively impacted only upstream and downstream neighbours. In such cases, it was much easier for the state to meet its duty to protect its citizens from harm through specific domestic regulatory measures or through bilateral or trilateral negotiations.

The increased mobility caused by globalization has shifted the concept of vicinity, so the prevention and mitigation of environmental risks are much more difficult. Such phenomena as the depletion of

---

[1] For details on the dimensions of globalization see Kjær 2004: 65–77; Held et al 2003: 387–412.

stratospheric ozone or anthropogenic global warming pose risks that are indeed worldwide. Because of their mobility and ubiquity, the problematic aspects of globalization can evade regulatory control by individual states.

The adverse effects of globalization are felt globally, while their cause often remains nebulous. According to Beck, this gives rise to a "world risk society" collectively affected by these risks (Beck 1997: 79; 2005: 22–9).

With the instruments it has at its disposal, the individual state cannot protect its citizens from global risks (Zürn 1998: 114–15). The enforcement of national law and the powers of an administration are generally limited to the state's territorial boundary. In view of the magnitude and complexity of global environmental problems and their harmful impacts on a population, the state is not able to perform adequately the key task of ensuring the security and welfare of its citizens. Therefore, globalization also entails the deterritorialization or transnationalization of regulatory areas (Beck 1997: 44–5; Hobe 1999b: 256; Hingst 2001: 112ff.).

Nevertheless, how can global risks be regulated and who is responsible for protecting the world risk society from the adverse effects of globalization given that there is no world government with the overarching authority to occupy these regulatory gaps and protect and preserve the welfare of the world population?

This gives rise to the question of what alternatives might be available to organize governance on a global scale (Rosenau 1992: 3). Political scientists have introduced the concept of global governance as a way of addressing global problems cooperatively. Global governance takes a universal approach to the resolution of global problems, aiming to reshape both institutional organizations and the attitudes of key actors; these include more or less anyone and anything able to contribute to a solution. While this approach might seem arbitrary, it is based on the idea that individual states and the traditional, state-based international order are not capable of tackling these issues on their own. Scholars emphasize here that the model of state-based governance is not exclusive; the implementation of other approaches is certainly conceivable (Id.: 4–5). This position recognizes that governance does not necessarily have to be backed up by a monopoly on the legitimate use of physical force, and that it is not always the state alone who acts for the common welfare.[2] Accordingly, the Commission on Global Governance defined

---

[2] Cf. Zürn 1998: 167–8 for several examples.

governance as:

> ...the sum of the many ways individuals and institutions, public and private, manage their common affairs. It is a continuing process through which conflicting or diverse interests may be accommodated and co-operative action may be taken. It includes formal institutions and regimes empowered to enforce compliance, as well as informal arrangements that people and institutions either have agreed to or perceive to be in their interest. (Commission on Global Governance 1995: 2)

Governance is disassociated from the state. The state may seek cooperative ways and exercise governance with civil society actors, *governance with the state*. Alternatively, civil society actors may act alone as a form of self-regulation, *governance without the state*, sometimes also described as *private governance* (Zürn 1998: 170f.). Global governance relies on governance contributions from a large variety of actors or coalitions of actors, allowing for several courses for action in order to address global problems.

As a result of the constraints that globalization imposes on a state's problem-solving capacities, the state is ultimately forced to relocate its governance resources to the transnational level (Scharpf 1991: 622). In order to work towards solving global problems, the state enters into "hybrid, multiparty, collaborative governance arrangements that pool and recombine the resources of a variety of state and non-state actors." These governance arrangements disaggregate and reassign powers that are usually exclusive to the sovereign state, pooling them with the powers, resources, and competences of other actors (Karkainnen 2002: 206–7). Because of their informal nature, these governance arrangements are aptly described as "transnational public governance." The use of the prefix "trans-" implies that these arrangements are concluded, and powers assigned, beyond the sphere of the nation-state, further distinguishing them from the highly formalized international domain.

As regards the institutional component of transnational public governance, bureaucracy networks have recently come to the attention of international lawyers (Slaughter 1997; Zaring 1998). These networks emerged as a result of transnational relationships between government officials, which have been observed since the 1970s (Keohane and Nye 1974; Tietje 2001). Closely connected to the emergence of networks is the role of law as an instrument of governance. Ostensibly, these networks are also involved in the creation of norms. In contexts of private governance, scholars have observed the emergence of transnational law. The primary example here is *lex mercatoria*, the private rules governing worldwide trade. These rules can overlap – sometimes even becoming substitutes for state-based law – a process described as *legal pluralism*: the coexistence of legal orders of different provenances governing similar subjects. Scholarly interpretations of this phenomenon are divergent. Some consider it to be a sign of the growing irreconcilable differences between legal orders (Fischer-Lescano and Teubner 2004); others view it as a globally linked system of legal rules – a system of *Interlegality* (Sousa Santos 2002). Here, the question is whether similar rules are also created in the context of transnational public governance and if and to what extent transnational bureaucracy networks contribute to the creation of a public version of transnational law.

Transnational public governance is not concerned only with institutions and instruments set up to solve global problems. It also must deal with problems stemming from its transnational and informal character. Scholars point out several problematic aspects of this type of governance, in particular, the "crisis of democracy" provoked by globalization (Scholte 2002: 289). The main reason for this crisis is the disassociation of the level where decisions are made and the level where they are implemented and individuals are affected. Traditionally, citizens elect a state government that then determines and implements particular measures that eventually bear upon the citizens. Global governance means that the measures affecting a constituency do not stem directly from the elected government, originating instead from institutional arrangements in which their government is one of many participants. The result is that decision-makers are not identifiable to the public, a circumstance that blurs accountability (Zürn 2004: 260). Consequently, territorially rooted mechanisms of democratic legitimacy do not effectively restrain transboundary governmental activities (Scholte 2002: 290). It is unclear in such cases whether governments are still clearly responsible for such developments (Kaiser 1998: 4).

Transnational public governance thus raises two kinds of questions. The first concerns its exact features. It has already been pointed out that bureaucracy networks and transnational law could be considered as possible institutions and instruments of transnational public governance. While the internationalization of national administrations and bureaucracy networks has already been subjects of several studies, an investigation of the legal implications is still missing. Transnational law not originating from private actors has also not yet been examined sufficiently. To address properly the question of legitimacy, it is important to determine how bureaucracy networks and transnational law are formed and operate and how they are interconnected with the formal legal order, and this makes the study of certain policy areas necessary.

Only once these insights have been gleaned from the practice of transnational public governance can the second set of questions concerning the legitimacy of bureaucracy networks and transnational law be tackled. This requires an assessment of the networks' actual significance, as well as of their possible interference with national legislation. On the basis of this knowledge, the question of legitimacy, as it arises in the context of transnational public governance, can be studied.

This book will attempt to address the legal aspects of transnational public governance and certain issues of legitimacy. To do this, it is necessary to first provide a general idea of the role of the state in the age of globalization. Some authors consider the growing role of non-state actors and the apparent impotence of the state as a sign of its diminishing role (Hobe 1999b: 269). Others take a radically different stance on the state's role, asserting that its powerlessness is a "myth," as globalization does not only constrain, but also empowers the state and its institutions (Weiss 1997; 1998: 188ff.; 2003a: 15ff.). The state has sufficient room to maneuver if it makes intelligent use of its assets. One possible means for the state to regain its effectiveness is through close interaction with other actors, such as other states or societal actors (Weiss 2003b: 298, 308–9). Part I will examine the role of the state and lay the groundwork for the rest of the study by outlining the current discussion on the ability of the state and the state-based system of international organizations (IOs) and international law to address global problems. It will focus on the current research on informal structures and governance in transnational settings. Of particular interest here are the recent findings concerning the structure of transnational networks of administrative bodies and the impact of transnational law.

Part II will explore public governance structures by examining a specific policy area in which the above-mentioned problems are prevalent: global efforts in the field of chemical safety. Chemicals pose a global problem; the usefulness of examining this case here lies in the facts that this has not yet been fully investigated and that few formal structures, such as international treaties, currently exist in this area. Part II will identify the relevant actors and the applied instruments, focusing on informal structures and rules. Further insights will be gained through the methodical analysis of relevant documents, such as memoranda of understanding (MoUs), terms of reference (ToRs), manuals, and other agreements, which will be supplemented by legal and toxicological literature. Since there are few documents concerning informal legal structures, information supplied in interviews will be relied upon to complete the overall picture. A legal sociologist and the author held guideline-based interviews with experts (*leitfadengestützte Experteninterviews*). Approaching the interviewees with an open set of guidelines instead of a fixed set of questions ensured that rich and relevant material could be collected (Meuser and Nagel 1991; 2006). The interviewees were officials in government agencies or IOs and scientists in private research institutions. They actively participated in these structures as delegates from national agencies, IOs, or scientific institutions and thus could provide deep insights into actual practices, which in some cases might deviate from the written rules laid down in the official documents. As the experts only agreed to be interviewed on condition of anonymity, no details regarding their occupation or employer can be provided.[3]

Part II will provide an overview of the factual situation on institutional and instrumental arrangements in the field of global chemical safety. The empirical findings will then be analysed and evaluated in Part III. The aim is to obtain a clearer picture of transnational public governance. From this analysis, it will be possible to identify the role of the state in international governance and the significance and problem-solving capacity of the state-based system, which involves IOs, public administration, and legal governance instruments.

Once a clearer picture of transnational public governance has emerged, the matter of its legitimacy can be addressed. If transnational public governance is a viable and practical method of solving global problems, the challenge will be to identify factors that can contribute to its legitimacy.

Weber has remarked that the bureaucracy is technically superior to other forms of organization because of its "precision, speed, unambiguity, knowledge of files, continuity, discretion, unity, strict subordination, reduction of friction and of material and personal costs." This does not mean that state bureaucracies do not occasionally pursue their own power interests; however, "in principle a system of rationally debatable 'reasons' stands behind every act of bureaucratic organization" (Weber 1978a: 973–9). Thus, it is generally assumed that administrative actions are based on "good" reasoning. But what criteria are

---

[3] On the interviewees' occupations and backgrounds, cf. the overview in the annex.

acceptable for justifying this reasoning if the administration operates in a transnational context? Part IV will attempt to determine such factors and investigate their application in transnational public governance arrangements.

This book will make extensive reference to the German legal system, particularly in the areas of constitutional and administrative law. Germany's constitution – the *Grundgesetz* – is relatively modern and has served as a model for the constitutions of other countries. An examination of the *Grundgesetz* may be of a broader international interest; however, the main purpose for using it here is its explicit mention of how the state is to conduct foreign affairs, a subject of several rulings of the Federal Constitutional Court. Collectively, the material on Germany's international relations, the transfer of sovereign rights to supranational organizations, and the exercise of authority in settings beyond the state is quite rich, especially since Germany's foreign policy has undergone several changes since the late 1980s. Therefore, despite the fact that this book relies in part on the distinctive characteristics of the German legal system and particular matters discussed here, such as the relationship between domestic and international legal orders, the exercise of foreign relations and the legitimacy of state activities exist in all Western states, and it will be possible to draw several general conclusions that are applicable to most Western legal orders. Occasional references to US constitutional and administrative law will also be made to illustrate certain crucial points.

Part I
Globalization and the State

2
The Changing State

Sovereignty is one of the defining elements of statehood (Grewe 2000: 166–7). It renders the state operational – both domestically and internationally. Domestically, the state holds the monopoly on the legitimate use of physical force and is legitimized and supported by the people (Weber 1978a: 56; Held et al. 2003: 45). Internationally, sovereignty determines the state's capability to participate in international affairs and to create and be subject to international law (Oeter 2002: 283; Morgenthau 1973: 309). In fact, modern international law rests on the concept of the sovereignty of states (Kimminich 1976: 97; Bleckmann 1995: 89ff.).

A state's sovereign existence is not an end in itself; the modern state is characterized by a number of national objectives (*Staatsziele*). While some of them are formal or structural principles – like democracy or federalism – others serve as orientation marks or optimizing requirements (*Optimierungsgebote*) for all state activities (Alexy 1994: 75–7). Welfare and security of citizens have always been basic state objectives. In modern constitutions, these are expressed as the rule of law, social welfare, peaceful international relations, and environmental protection.[4]

Globalization affects the performance of state functions. For example, the concept of citizenship is loosing significance (Kokott 2003: 12–13). Governance is no longer confined to a specific territory, now having extraterritorial reach. Territorial borders – originally delimiting a state's jurisdiction – have become increasingly permeable, for example, with the exterritorial application of antitrust laws (Id.: 13–14; Herdegen 2003: 38ff.). On the other hand, governments may prove to be incapable of preventing the effects of harmful substances on their territory and

---

[4] Cf. Sommermann 1997: 198–252 for examples from several constitutions.

population; and the sources of many global environmental problems continue to evade the regulatory grasp of national governments.

Globalization has also influenced the incorporation of international elements into state policy objectives (Sommermann 1997: 252–3). Environmental protection, aimed at preserving human and other life, cannot be maintained by domestic measures alone. This means that the state is no longer able to guarantee the security and welfare of its citizens on its own. The pursuit of policy objectives has become an international matter, where obligations are owed to the international community as a whole, and not solely to the citizens of a nation state (Tomuschat 1999: 94–5).

Solving global problems is inextricably linked with the performance of state functions. Consequently, the state is forced to seek ways to cooperate with other states and non-state actors. This chapter investigates the question of how the state adapts to this necessity and what is necessary for fruitful cooperation between states.

Sovereignty and international cooperation

Sovereignty determines a state's identity and defines its relations with the rest of the world. For example, if a state assumes a realist approach to sovereignty, it might conceive of its sovereignty as an impenetrable sphere, shielding it from the influence of other states and regarding it as something indivisible.[5] Entering into cooperative relations with other states or transferring sovereign rights to a higher authority is difficult from this perspective. Hence, a state's concept of its own sovereignty determines its openness and willingness to cooperate with other states, its willingness to engage itself in governance beyond its territory and its acceptance of external sources of governance within its own territory.

Although sovereignty is a dynamic concept, dependant on the ever-changing context of history, its key characteristics have prevailed over time (Perrez 2000: 244–7; Schrijver 1990: 70). Sovereignty, as conceived in the sixteenth century by the French political philosopher *Jean Bodin* or implemented in the Peace Treaties of Westphalia of 1648, has undergone many changes and differs considerably from today's understanding (Fassbender 1998: 26–33). The growing interdependence of states has accelerated these changes.

During the sixteenth and seventeenth centuries sovereignty was outlined as a core element of the modern, post-medieval state. Without

---

[5] Morgenthau 1973: 319–24 conceived of sovereignty in this way.

using the term "sovereignty," *Bodin* outlined the concept in his seminal work, *Les six livres de La République* (*The Six Books of the Commonwealth*), as the "absolute and perpetual power of the Commonwealth." The power of the prince is absolute, restricted only by the laws of God and natural law (Bodin 1981: 205–10). *Hugo Grotius* regarded sovereignty in a similar way, viewing it as the supreme authority, subject to nothing else except the will of its bearer (Grotius 1950: 93–4).

The Peace Treaties of Westphalia implemented this idea of sovereignty (Perrez 2000: 19–25; Fassbender 1998). Autonomous and independent coexistence characterized the new international order instituted under the Peace of Westphalia (Gross 1948: 28–9).

The unity of a country's government and the impenetrability of its outer sphere mark the Westphalian state. Internally, the state is independent from other powers: the ultimate authority with a general, all-encompassing competence, the monopoly on lawmaking and the monopoly on the legitimate application of force to implement its laws. Externally, this independence precludes interference in state affairs by other powers. Other states are denied the possibility of intervening in another state's internal affairs. This concept of sovereignty prevailed for as long as states could satisfy the needs and common interests of their population in an almost autarkic way. International law had the purpose of regulating the non-interference of states in each other's affairs and providing for their peaceful coexistence (Bleckmann 1995: 738). In its 1927 judgement of the *S. S. Lotus* case, the Permanent Court of International Justice defined the role of international law in regard to sovereign states:

> [i]nternational law governs the relations of independent States. The rules of law binding upon States therefore emanate from their own free will as expressed in conventions or by usages generally accepted as expressing principles of law and establishing in order to regulate the relations between the co-existing independent communities or with a view to the achievement of common aims. Restrictions upon the independence of states cannot therefore be presumed.[6]

Under this classical concept of the relationship between sovereignty and international law, cooperation among states is restricted to diplomatic

---

[6] Permanent Court of International Justice, Judgment of 7 September 1927: 18.

collaboration. As such, international law as a legal order was rather ineffective, since states relied instead on their power to protect or enforce their interests (Bleckmann 1995: 759–61).

Under the impression of the failure of the League of Nations and with the onset of a new era of international institutionalized cooperation after World War II, the academic concept of sovereignty changed. For example, *Kelsen* defined sovereignty "...as the legal authority of the states under the authority of international law" (Kelsen 1944: 208). International law assumed a dual position: restricting sovereignty and guaranteeing it at the same time by protecting states from illicit interference into their affairs by other states (Perrez 2000: 48). The adoption of the United Nations (UN) Charter reflects this profound systemic change. During the 1940s and 1950s, the term "cooperation" found its way into legal documents. Prominent in this regard is Chapter IX of the UN Charter, which is entirely devoted to "international economic and social co-operation" (Loewenstein 1954: 224–5). *Friedmann* showed in the 1960s how this resulted in structural changes to international law, towards ushering in a new concept of international law based on cooperation. He distinguished it from the classical international law of coexistence, which traditionally regulated the peaceful coexistence of states through rules governing diplomatic relations where mutual respect for state sovereignty was the governing principle (Friedmann 1964: 60).

Today, the UN Charter provides the most general basis for international cooperation (Arts 1 (3), 11 (1), 13 (1), 55 (a) and (b), 56) (Perrez 2000: 268ff.; Cassese 150ff.). Two resolutions of the United Nations General Assembly (UNGA) have concretized the UN Charter's relatively abstract provisions: the Declaration on Principles of International Law concerning Friendly Relations and Co-operation among States in Accordance with the Charter of the United Nations (UNGA Res. 2625 (XXV)) of 1970[7] and the Charter of Economic Rights and Duties of States (UNGA Res. 3281 (XXIX)) of 1974. The Declaration proclaims as its third principle "[t]he duty of states to co-operate with one another in accordance with the Charter [of the United Nations]." In the same vein, Art. 9 of the Charter sets out a "...responsibility to cooperate in the economic, social, cultural, scientific and technological fields for the promotion of economic and social progress throughout the world, especially the developing countries." The matter is underscored by Art. 17 of the Charter, where "[i]nternational co-operation for development is the shared goal and common duty of all States." Of course, despite the

---

[7] Cf. Verdross/Simma 1984: §505 for a discussion of the Resolution.

treaty-like language in parts of both resolutions, they are, like all UNGA resolutions, non-binding.[8] Moreover, the at times cautious phrasing reflects the smouldering East-West-conflict of the 1970s.[9] However, these resolutions certainly play an important role in the formation of a rule of customary international law (Tietje 2001: 225). Other soft law instruments have also included provisions on international cooperation, for example, as stated in the Final Act of the Conference on Security and Co-operation in Europe (the CSCE Final Act of 1975) in Art IX: "The participating States will develop their co-operation with one another and with all States in all fields in accordance with the purposes and principles of the Charter of the United Nations."

Similar provisions can be found in international instruments concerning environmental issues. For example, Principle 24 of the 1972 Declaration of the United Nations Conference on the Human Environment (hereafter referred to as the Stockholm Declaration) declares that "[i]nternational matters concerning the protection and improvement of the environment should be handled in a cooperative spirit by all countries, big and small, on an equal footing." The 1992 Rio Declaration on Environment and Development and the related Agenda 21 pick up this language. Principle 12 of the Rio-Declaration calls upon states to "...cooperate to promote a supportive and open international economic system that would lead to economic growth and sustainable development in all countries, to better address the problems of environmental degradation." Agenda 21 as a whole is a blueprint for a joint effort not only by states but also of non-state actors to ensure a sustainable development. Several treaties have concretized these formulations and states commit to address an environmental problem jointly (Stoll 1996: 54; Perrez 2000: 304–30). The elements of cooperation are rather sophisticated, ranging from the exchange of information through notification and consultation to close collaboration in administrative, financial, scientific, and technical matters (Stoll 1996: 64–81).

The development of international environmental law during the twentieth century – itself propelled by the rise in transboundary pollution caused by industrialization – had a great impact on the modification of concepts of sovereignty (Hinds 1997). For example, international jurisdiction responded to the tension between sovereignty and transboundary pollution. The rationale supporting the landmark *Trail Smelter*

---

[8] For an analysis of the Charter against the background of the legal character of UNGA resolutions cf. Tomuschat 1976: 465–90.

[9] For the Declaration cf. the analysis by Rosenstock 1971.

decision was based on the international legal principle of *sic utere tuo ut alienum non laedas* – use your property so as not to harm that of others. When deciding the case of a smelter on Canadian soil emitting fumes which caused damages in the United States, the arbitral tribunal ruled in 1941 that under the principles of international law "... no State has the right to use or permit the use of its territory in such a manner as to cause injury by fumes in or to the territory of another or the properties or persons therein, when the case is of serious consequence and the injury is established by clear and convincing evidence."[10] This principle was later affirmed in the *Corfu Channel*[11] and *Lac Lanoux*[12] cases and further acknowledged upon its inclusion into the Stockholm Declaration as Principle 21. States can exploit resources within their territory, as long as this does not lead to environmental damage in other states (Wolfrum 1990: 309–18). This principle requires that states respect the sovereignty of other states, thereby limiting their own sovereign rights. Ultimately, this tension can only be resolved if states choose to coordinate their activities and eventually cooperate in the area of environmental protection.

A specific consequence of this principle is a duty on states to report transboundary impacts to affected states (Stoll 1996: 47; Hinds 1997: 145–80; 259–80; 329–42). This duty is not confined to cases in which the damage has already occurred or is underway, but also when, for example, in the course of environmental planning, an impact might be expected. Affected states are to be given the opportunity to voice and discuss their concerns (Stoll 1996: 48–9). This procedure primarily serves the purpose of reconciling the sovereign interests of the involved countries, and does not serve a higher, common interest (Id.: 50); but close interstate cooperation to address common problems jointly is rooted in this principle.

---

[10] Reports of International Arbitral Awards, Trail Smelter Case, 1965.

[11] International Court of Justice (ICJ), Judgement of 9 April 1949, Corfu Channel Case, 22: "Such   obligations [notifying foreign ships of minefields in territorial waters] are based ... on every State's obligation not to allow knowingly its territory to be used for acts contrary to the rights of other States."

[12] Report of International Arbitral Awards, Affaire du Lac Lanoux, (19 November 1956), p. 315: "The Tribunal is of the opinion that, according to the rules of good faith, the upstream State is under the obligation to take into consideration the various interests involved, to seek to give them every satisfaction compatible with the pursuit of its own interests, and to show that in this regard it is genuinely concerned to reconcile the interests of the other riparian State with its own."

In the end, the promotion of cooperation in international law has given rise to a general duty of states to cooperate. In the mid-eighteenth century, the Swiss jurist *Emer de Vattel* construed a duty to cooperate as a legal principle in his work on international and natural law. The goal of the international community of states is to assist each other in the course of improving and perfecting their condition (de Vattel 1959: 22). Today, the UNGA resolutions pertaining to international cooperation are understood to be *opinio juris* (opinions of law), expressing a general duty of states to cooperate as a principle of international customary law. Principles of international law always require further specification, as they are too abstract to impose any explicit obligations on states. Nevertheless, the purpose of interstate cooperation can be clarified. Art. 1 (3) and Arts 55 and 56 of the UN Charter define the ultimate purpose of international cooperation: the promotion of peace, human rights and international welfare (Tietje 2001: 226–32).

The developments mentioned above can be regarded as restrictions on state sovereignty. However, the last process to be outlined here concerns the bundling of sovereign rights on the inter- or supranational level. When states create International Organizations (IOs), they exercise their sovereignty, but at the same time restrict it when they simultaneously transfer sovereign rights to the newly founded organization (Delbrück 2001: 13; Sassen 1996: 29–30). Transferring sovereign rights means that states relinquish a part of their all-embracing authority to the newly created IOs.

While states are very cautious when relinquishing sovereign rights and transferring powers to an international entity, there has been an emergence of powerful organizations at a regional level. Member states of the European Union (EU) have ceded sovereignty to a superordinate institution more than any other states in the world. The EU is endowed with a vast array of competences. However, the principle of conferred powers (*Prinzip der begrenzten Einzelermächtigung, competences d'attribution*) prevents a universal and general competence (Hartley 2003: 105ff.; Arnull et al. 2000: 153ff.). Nevertheless, European law overarches domestic legal orders.[13] Notwithstanding the fact that the member states are still *"Herren der Verträge"* – masters of the treaties,[14]

---

[13] European Court of Justice (ECJ), Judgement of the Court of 15 July 1964 (Costa v. E.N.E.L: 593).

[14] German Federal Constitutional Court, Order of 8 April 1987: 242; Federal Constitutional Court, Judgement of 12 October 1993: 190; cf. Hartley 2003: 157–60 for a brief overview of the Federal Constitutional Court's famous Maastricht-decision.

they have transferred sovereign rights to the EU, creating a supranational order that affects not only them, but also their citizens.[15] As far as the members of the EU are concerned, their authority has been eroded to a significant degree (Hobe 1998b: 369–79). Instead, it has been put into the service of European integration.[16] European states have responded to their inability to master transboundary problems by creating a supranational order. Certain rights emanating from sovereignty were transferred and bundled at the supranational level in order to create a capable cooperative institution – a process that is a reflection of the development that ultimately lead to the centralization of supreme authority in the modern sovereign state. In doing this, individual states have retained areas of policy in which to act or make decisions autonomously (Id.: 435–6).

However, it is clear that the development of the EU is owed largely to recent events in European history, the specific regional economic necessities, and the relative homogeneity of European culture.[17] As a result, similar developments in other regions of the world or at a global level are unlikely to occur in the near future. The transforming idea of sovereignty is also reflected in certain constitutional changes, where national constitutions have been adapted to facilitate the incorporation of international law. A strict dualism, as described by *Triepel* (Triepel 1899), is no longer upheld: a strict separation of the national and international legal orders is artificial and cannot be maintained in an interdependent world, necessitating of international cooperation between states (Bleckmann 1995: 764–5). As a consequence, constitutions have been designed to allow for permeability of the state's legal sphere; and this openness towards cooperation is evident in several modern constitutions.[18]

The German *Grundgesetz* (Basic Law) contains provisions integrating Germany into the international legal order.[19] Initially, however, it adopts a dualistic stance, viewing the national and the international legal orders as separate legal systems. International law may be applied

---

[15] ECJ, Opinion of the Court 14 December 1991: 6102, para. 21; ECJ, Judgement of the Court 5 February 1963: 13.

[16] Federal Constitutional Court, Judgement of 12 October 1993: 188–9.

[17] Nevertheless, it is still difficult to identify a European *demos*, cf. *infra* 193.

[18] Cf. Häberle 1978: 149 and 167ff. – for the world of 1977. However, his observation remains applicable, it is very likely that even more countries, especially the post-socialist ones, which transformed their form of government or emerged in the 1990s, adopted "open" provisions.

[19] Cf. Federal Constitutional Court, Order of 22 March 1983: 370.

within the state only after it has been transformed (Hillgruber 2004: marginal n. 116). Pursuant to Art. 59 II 1 of the *Grundgesetz* international treaties within the meaning of Art. 38 (1) (a) of the ICJ-Statute the parliament is required to enact a law in order to ratify an international treaty. Accordingly, international treaties enjoy the same status as any ordinary Act of Parliament (*einfaches Bundesgesetz*). Art. 25 of the *Grundgesetz* breaks with this dualistic stance insofar as it directly incorporates universal international customary law within the meaning of Art. 38 (1) (b) and (c) of the International Court of Justice (ICJ)-Statute into the national legal order.[20]

Provisions allowing or even prescribing participation in institutionalized forms of cooperation supplement the general permeability of a state's constitutional and legal order to international cooperative efforts (Hobe 1998b: 423). According to Art. 23 of the *Grundgesetz*, Germany shall participate in the development of the EU. Art. 24 para. 1 *Grundgesetz* allows the transfer of sovereign rights to intergovernmental organizations, provided the transfer is based on an international treaty and has the approval of the legislative branch of the state. Furthermore, Art. 24 para. 2 of the *Grundgesetz* enables the federal government to accede to collective security systems, that is the North Atlantic Treaty Organization (NATO), and accept certain restrictions of its sovereign rights. The transfer of sovereignty also finds its limits in the *Grundgesetz*. The Federal Republic of Germany must not be absorbed by the supranational order and the core of the basic rights guaranteed in the *Grundgesetz* or the federal structure must remain untouched.[21]

In sum, the German *Grundgesetz* allows for Germany's integration into the international community, formulating an obligation to participate actively in international affairs (Vogel 1964: 46). The capacity to solve problems or to attend to certain issues is not necessarily concentrated at the state level. Instead, problem-solving capacities can be allocated to the level at which they can be used most effectively. Competences and resources directed at the resolution of problematic issues can be transferred to local, federal, regional, international, and transnational levels (Hobe 1998a: 531; 1998b: 392, 419ff.), while the *Grundgesetz* not only allows for but also limits the transfer of sovereign rights to supranational or international institutions (Mosler 1992: marginal ns 63–79).

---

[20] Federal Constitutional Court, Order of 30 October 1962: 32ff.; Order of 9 June 1971: 177.

[21] Federal Constitutional Court, Order of 22 October 1986: 375–6; Cf. also Mosler 1992: §175, marginal ns 65–75.

Similar provisions can be found in the constitution of the United States. Under Art. II Section 2 clause 2 and Art. VI clause 2 of the US Constitution, a treaty is concluded by the President and approved by a two-thirds majority of the Senate, it becomes law. Non-self-executing treaties, however, require an explicit Congressional act of implementation (Henkin 1997: 198–206).

Although not as explicit as the German *Grundgesetz*, the US Constitution also allows the President to transfer authority to IOs. The constitutional provisions on international treaties and executive agreements provide sufficient basis for accession to IOs. Sovereign rights may be transferred, although this process must not be irrevocable (Id.: 247–66).

In general, one can observe an overall openness – with varying degrees of divergence and some reservations – on the part of modern constitutions towards international law and cooperation (Cassese 1985). *Häberle* coined the term "cooperative constitutional state" to describe the state in which public authority is bound by constitutional provisions, thus guaranteeing an open and pluralistic society that is actively involved in the concerns of foreign states and their citizenry (Häberle 1978: 144–5).

Sovereignty, as the key feature of the state, has undergone many changes since its inception; the most profound are the result of the increasing interaction and interdependence of states during the twentieth century. In view of the current challenges – most importantly global environmental problems – it is clear that the conception of sovereignty as a shield or impenetrable sphere can no longer exist. While the state can isolate itself legally or politically, it cannot evade the impact of these problems (Schrijver 1990: 97) nor can it completely surrender sovereignty. Yet is the cooperative state of the early 21sh century prepared to face the challenges of globalization?

A recent debate among international lawyers focused on the significance of sovereignty in a globalized world with highly interconnected and interdependent states. Some speak of an "erosion" of sovereignty (van Staden/Vollaard 2002), questioning whether the concept of sovereignty is outdated (Oeter 2002), while others focus on the changes of the concept to reflect a "new sovereignty" (Chayes/Chayes 1995) or "cooperative sovereignty" (Perrez 2000).

Some international legal scholars find that the restrictions imposed on sovereignty by international court rulings and the transfer of sovereign rights to IOs are inadequate when it comes to dealing with global environmental problems. The balancing of territorial authority and territorial integrity as one of the fundaments of international environmental protection has its limits (Odendahl 1998: 301–13). Restricting sovereignty is adequate only when dealing with transboundary pollution that clearly relates to a specific emission source. This notion of sovereignty regards the state as being wholly independent. It is also founded on the assumption that state actions have a local impact, while in reality, the effects of such actions may be felt globally, and that the availability of natural resources exceeds consumption, when in fact resources are depleted rapidly, and thus reality contradicts these assumptions (Perrez 2000: 114ff. and 169ff.).

To remedy this problem, these scholars suggest a reshaping of sovereignty: instead of allowing activities harmful to the environment to proceed, territorial sovereignty must be understood as a state responsibility to protect the environment within its territory. This duty would be similar to the state's responsibility to its citizens or its cultural heritage, which also emanate from its sovereignty (Odendahl 1998: 362–70). The sovereignty paradigm must change from sovereignty as independence to an "enlightened" sovereignty, which incorporates the wellbeing of mankind into state interests (Hobe 1998b: 281). Today, interdependence encompasses not only the linkages between states, but also those with non-state actors. In order to maintain a certain degree of ability to act, states must adapt their concept of sovereignty. They can exercise their sovereignty only through participation and cooperation in global regimes. Reluctant or uncooperative states lose allies and the ability to participate in effective coalitions with others, thereby limiting their room to manoeuvre, which cannot be regained by acting independently and autonomously. Ultimately, sovereignty is equivalent to the state's standing or reputation in the international system (Chayes/Chayes 1995: 27).

The acknowledgement of "universal values" in the international community of states indicates an ongoing process of change. The international protection of human rights is one of the factors that triggered a recent change in the legal concept of sovereignty. Protecting individuals is not only a state concern, as the international community itself has assumed this responsibility, thereby encroaching on the national sovereignty of individual states (Wolfrum 1990: 308; Schrijver 1990: 83). Furthermore, states have repeatedly taken action to preserve the environment in areas beyond the reach of national jurisdiction, such as Antarctica or the High Seas, thereby recognizing the global environment as another "universal value." Global environmental protection, however, requires that measures be taken on a state's own territory, necessitating further infringements of its sovereignty (Wolfrum 1990: 327).

Thus far, globalization has not deprived states of their sovereignty. It has only been operational sovereignty, the ability to actually exercise sovereign rights that has been obstructed (Reinicke/Witte 2000: 81). Moreover, the commitment to sovereignty has not been abandoned, rather, it has been placed into the service of the common welfare of the international community (Dahm/Delbrück/Wolfrum 1989: 222; Tomuschat 1999: 262–3). Eventually, the question arises of how much of its sovereignty a state can transfer or how severely it can be restricted before sovereignty is no longer a constitutive feature of the state that represents its monopoly on force, exclusive jurisdiction, and ultimate responsibility.

Administration in the age of globalization

A public administration exists as an instrument executing the will of the state, which has been previously defined through political processes. According to *Weber*, public administration is a precision instrument for the exercise of authority, because its operation is strictly hierarchical, based on discipline and obedience (Weber 1978a: 220–1). In order to fulfil its functions properly, the administration requires leeway, which is granted by the legislator in the form of discretion. With increasingly complicated administrative tasks, the scope of this licence has grown. Lawyers observed in the 1970s that an increasing number of laws contained purposive wording that created programmes by setting out in the statute an objective to be achieved by the administration, instead of employing conditional phrasing, that is, telling the administration exactly how to act under specified circumstances (Schmidt 1971). In the case of conditional programmes, the legislator has ultimate control on the output of administrative activities. Purposive programming involves a loosening of control. This, however, entails the danger that the administration will use its autonomy to develop its own political ideas – outside parliamentary or ministerial lines of control (Mayntz 1985: 66–7; Schmitt Glaeser 1973: 203).

Globalization affects the way in which a national administration performs its functions. The following section will take a look at administrative practice in the age of globalization. It will be necessary to determine if the administration has become so autonomous that it operates outside the effective control of its superiors, notably Parliament, and if so, whether there are ways to re-establish control.

Traditionally, the formulation and exercise of foreign policy is the task of the ministry in charge of foreign affairs. In the 1920s, despite the fact that states had already been engaged in international cooperative measures like administrative unions (AUs) for several decades, the foreign ministry was thought to be the only point of access for inter-state communication and cooperation (Wolgast 1923: 78). German lawyers traditionally considered the administration and its activities to be deeply linked with the national legal order and the state territory. The German scholar of administrative law *Mayer* remarked: "[a]dministration is an activity of the state, performed under its legal order" (Mayer 1924: 9).[22] Fostering foreign relations was regarded as too specific to form a part of the general administration (Forsthoff 1973: 14).

Today, almost every regulatory matter has an international component. Thus, it is no longer possible for foreign ministries to uphold their claim as the sole representative of the state in foreign affairs (König 2000: 500–1). The primacy of foreign and security policy over other policy areas is no longer sustainable, as the security concerns of the state become multidimensional and transboundary issues permeate virtually every policy area (Eberwein/Kaiser 1998: 3). Close interstate coordination and cooperation require the involvement of the competent specialized ministries (Bleckmann 1995: 766); the foreign ministry retains its core functions – consular services and the fostering of diplomatic relations – while the relevant ministries and agencies liaise with their counterparts in other countries and work towards the solution of common problems. Thus, there is a blurring of the line between domestic and foreign affairs (König 2000: 501; Cassese 1985: 60).

With regard to Germany, which may serve as an example for state practice, at least in Organisation for Economic Cooperation and Development (OECD) countries, foreign affairs have been subdivided into two areas: of foreign policy and foreign relations. *Foreign policy* is defined by the overall interests of the state and determined and carried out by the Chancellor, the cabinet and the foreign ministry. *Foreign relations* are maintained by competent ministries and authorities, and are not necessarily congruent with the states' foreign policy.

According to the federal government's rules of procedure (RoP) and the federal ministries' joint rules of procedure, the foreign ministry has a supreme role in foreign affairs. §11 of the *Geschäftsordnung der Bundesregierung* (*GO BReg*, Rules of Procedure of the Federal Government) stipulates that receiving foreign or international delegations and the

---

[22] Translation by the author.

commencement of negotiations with or in foreign countries requires the consent, and, when required, the participation of the foreign ministry. According to §38 of the *Gemeinsame Geschäftsordnung der Bundesministerien* (Joint Rules of Procedure of the Federal Ministries), supreme federal authorities may cooperate with organs or other subunits of foreign countries or IOs only if cooperation is based on an international or bilateral agreement or if the foreign ministry or the federal ministry consented to the cooperation. There is also a legal framework for the international activities of other ministries and agencies, but the example of coordination and cooperation in environmental affairs shows that the necessary internal coordination processes between the relevant actors of foreign environmental policy is in fact a lot less formal (Fischer/Holtrup 1998: 126). This gives the state to the challenge of monitoring the various activities of its subunits, orchestrating its administrative subunits to yield maximum results and maintaining consistent representation through avoiding contradictions and solo attempts (König 2000: 501ff.; Cassese 1985: 61–8).

The state employs its administration to perform its two key functions: safeguarding the security of its citizens and advancing common welfare. It has already been pointed out that globalization extends to national policy objectives; this means that it is very difficult for a national administration to pursue these objectives on its own. Globalization affects the capability of the state and its administration in such a way that its functions can no longer be performed effectively (König 2000: 478) and broadens the public administration's area of responsibility (Farazmand 1999: 519) while its competencies remain unchanged. Thus, in the same way that regulatory items transcend national borders, national agencies must also operate internationally (König 2000: 478–9).

Constitutional provisions complement this necessity. Art. 24 of the *Grundgesetz* should be interpreted as a policy objective, laying down the aim of international cooperation (Tietje 2001: 216–17). From a constitutional perspective, the administration is obliged to enter into transnational administrative cooperation (Id.: 235ff.) and consequently, the national administration undergoes a process of internationalization (Delbrück 1987: 388). In the various forms of cooperation at the international level – through AUs, IOs and probably most significantly in the shape of the EU – national administrations form links with each other, bringing about a *deterritorialization* or territorial extension of administrative competencies (Tietje 2001: 179ff.).

The United States administrative system has also undergone a process of internationalization. Independent agencies such as the

Environmental Protection Agency are one of the striking aspects of the United States administrative structure and increasingly these agencies act on the international stage, cooperating with their foreign counterparts and concluding international agreements, sometimes without the awareness of Congress or the State Department. Because of their design, they enjoy greater independence and autonomy than their German counterparts. Their international activities may even impinge on the President's competences as Chief Executive and sole organ of the federal government in the field of international relations (Henkin 1997: 129–30).

A process of *disaggregation* accompanies the internationalization of the administration. The modern state, with its sectoral and – at least in the federal state – territorial division of labour, together with the processes of decentralization and deconcentration, provides administrative bodies with the resources and autonomy to communicate and collaborate with their foreign counterparts (König 2000: 489). The individual state's powers are destabilized by globalization; to the norm of centralized power at government level is outdated and ineffective and instead, new forms of governance emerge, involving the subunits of the state (Picciotto 1997: 261).

The operation of administration at the transnational level seems to be a logical consequence of the open, cooperative state. Yetthis raises the question of how such an administration affects the state. Does transnational administrative cooperation result in the end of state sovereignty? The examples of the German *Geschäftsordnung der Bundesregierung* and the *Gemeinsame Geschäftsordnung der Bundesministerien* show in principle that international activities are coordinated and controlled by the Foreign Office. But can such a degree of control always be maintained? Most importantly, to what extent are parliaments, which are in a representative democracy the ultimate source of legitimacy for all executive measures, involved in these transnational processes? One objective of the Part II is to provide sufficient data to attempt to answer such questions.

3
Civil Society Actors

The sovereign state has become the "prototype of [the] international actor" (Schreuer 1993: 448). States create international law and international organizations (IOs). Other stakeholders affected by legal agreements at the international level are generally not admitted to these processes. Nevertheless, civil society actors have emerged in the past 150 years, representing certain groups and interests, aiming to influence international policies. In the mid-nineteenth century, nongovernmental actors began to form associations across national borders. Probably the most famous of these are the International Committee of the Red Cross and the International Movement of the Red Cross and the Red Crescent, founded in 1863 by Henri Dunant and notable citizens of Geneva (Amerasinghe 2005: 3; Klabbers 2002: 17–18).

Types of civil society actors

Today, a large number of civil society actors operate globally – further evidence of the globalization process. There are two types of civil society actors: Nongovernmental Organizations (NGOs) and Transnational Enterprises (TNEs).

There is no generally accepted definition of NGOs, but the term commonly includes independent, permanent, organized, non-profit, nongovernmental associations that operate internationally (Ipsen 2004: § 6, marginal n. 19). NGOs can be further distinguished from liberation movements, since NGOs do not aim to overthrow governments. While NGOs criticize governments and are involved the political process, they are not political parties, since they do not aspire to seize state power. They raise funds to carry out their activities, but do not aim to make a profit, distinguishing them from corporations. Business associations such as the

American Chemistry Council (ACC) or *Conseil Européen des Fédérations de l'Industrie Chimique* (CEFIC), which represent industrial interests beyond an individual company should also be included in this category. These lobby for industry-friendly decisions, and do not operate for a profit either. Finally, their disrespect for the legal order may amount to nothing more than civil disobedience, distinguishing them from criminal organizations (Kamminga 2002: 390; Dahm, Delbrück, and Wolfrum 1989: 232ff.).

TNEs share with NGOs the characteristics that they are not established or maintained by states and operate across borders. However, TNEs and NGOs can be distinguished on the basis of their objectives. While NGOs pursue a public purpose, TNEs aim for profit (Thürer 1999: 46–7). Thus, TNEs can be understood as organized units that are established in more than one state (OECD 2000: 17–18). The transnationalization of corporate structures reflects the need for the division of labour, specialization, and operation in various national markets to remain competitive (Herdegen 2003: § 3, marginal n. 38).

Civil society actors in the international system

From a policy perspective, the role of civil society actors differs immensely. Generally, NGOs represent societal interests and aim at the implementation of certain values in the international system. They represent a wide spectrum of interests. For example, Greenpeace International tries to raise awareness of environmental issues through spectacular campaigns, thereby exerting pressure on both governments and corporations. Amnesty International promotes human rights, the Cooperative for American Relief Everywhere provides relief in disaster areas, and Médecins Sans Frontières offers health care and medical training in undeveloped countries and conflict zones. Organizations like the International Federation of Pharmaceutical Manufacturers and Associations or the International Federation of Agricultural Producers represent business and industry interests, acting as spokespersons before IOs like the World Trade Organization (WTO).

The revenue of the world's largest corporations exceeds the gross domestic product of most nations.[23] Of course, the economic power

---

[23] According to *Fortune* (2008), Wal-Mart Stores generated a revenue of US $378,799 million, making it by this measure the largest corporation in the world. The company operates in 200 countries. An estimate for Pakistan's GDP based on purchase-power parity amounts to ca. US $378,225 million, ranking it 24th in a list of 163 national economies (cf. World Bank 2007).

expressed in these numbers does not actually rival the powers of sovereign states (Wallace 2002: 66ff.). But with production facilities, business activities and personnel from all over the world, these companies operate transnationally, exerting immense influence over nations across several dimensions. A decision to close facilities in a certain part of the world can reverberate through an entire national economy. For better or worse, corporate governance can have direct effects on a country's labour and safety standards, education, and welfare systems (Muchlinski 1997: 90ff.).

Differing objectives notwithstanding, it is clear that NGOs and TNEs have both amassed significant powers, making them cornerstones of the architecture of global governance. The question now is whether and, if so, how they are integrated into international law, the legal underpinning of global governance.

NGOs and TNEs are both legal entities under national laws. International law, however, does not recognize either as such. On the international stage, the state is the primary actor. IOs derive their status as subjects of international law from the constitutive acts of states. The importance of their role within the international system is acknowledged, whereas civil society actors are more or less neglected. However, it is not unprecedented for states to accord civil society actors with subjecthood under international law. The cases where this has happened are relatively obscure, however.[24] Academic discussion already postulates a legal status for NGOs and TNEs under international law (Delbrück 2002a: 411–14; Hobe 1997).

NGOs are usually awarded only consultative status. The UN Charter recognizes the importance of NGOs in Art. 71, which authorizes the Economic and Social Council (ECOSOC) to arrange for consultation with NGOs. Obtaining consultative status endows NGOs with certain procedural rights (ECOSOC Resolution 1996/31). In this way, only the procedural rules, which constitute secondary rules of international law, recognize NGOs (Hobe 1999a: 171). While provisions like ECOSOC Resolution 1996/31 para. 50 invite NGOs to express their views and concerns at the international level, they are usually precluded from actively participating in the negotiation of international treaties (Kamminga 2002: 393ff.). However, the negotiation of several international treaties has been actively followed by NGOs and some treaties acknowledge

---

[24] Cf. Dahm/Delbrück/Wolfrum 1989: 317–38 for an account of the Holy See, the International Committee of the Red Cross, and the Sovereign Military Order of Malta as traditional subjects of international law.

their role by integrating them in implementation or follow-up proceedings (Id.: 394–9; Suy 2002: 376–80). Recently, the primary rules
of international law have also provided observer status to NGOs. Some
conventions have even accorded legal status to NGOs (Delbrück 2001:
25; Dahm, Delbrück, and Wolfrum 1989: 241–2).

Despite their involvement in the creation and implementation of
international law, NGOs still do not enjoy the status of international
legal entities (Id.: 243; Suy 2002: 385). Their formal status within the
international legal order remains weak and its development depends
largely upon the interests and preferences of sovereign states (Kamminga
2002: 404), which still exclusively decide on international matters (Suy
2002: 385).

Notwithstanding their weak legal status, NGOs play an important
role in the international legal system. Via their own mechanisms, NGOs
can initiate lawmaking processes and ensure compliance with international law, thereby enhancing its performance, often by denouncing
noncompliance or exposing free riders (Chayes and Chayes 1995: 251).
They influence public opinion, exerting pressure on states, IOs, or other
civil actors such as TNEs. Judge Weeramantry acknowledged their influence on international courts when he pointed out the legal relevance
of public opinion expressed in the NGOs' submissions to the court.[25]
Similarly, the Appellate Body acknowledged the role of NGOs, when
it ruled that it could consider their *amicus curiae* (friend of the court)
briefs (Ohlhoff 1999: 141ff.).[26] Due to these functions, NGOs purport-
edly have the capability to strengthen the international system, especially the lawmaking process. They attend international conferences,
interact with state delegates, and disseminate their views, acting as the
voice of the civil society. They also help to legitimize and increase the
quality of international decision-making, providing expertise in their
field of activity (Delbrück 2001: 18–19). Yet, critics perceive the role and
influence of NGOs as disproportionate, because they are virtually unac-
countable. In fact, some consider them to be a threat to the Westphalian
system (Kamminga 2002: 388).

TNEs are not considered to be subjects of international law either
(Dahm, Delbrück, and Wolfrum 1989: 243–58). However, their international importance has attracted the attention of various IOs, which have
addressed in codes of conduct the roles of TNEs. The most prominent

---

[25] ICJ, Advisory Opinion of 8 July 1996 – Dissenting Opinion of Judge
Weeramantry, 533–4.

[26] WTO, Report of the Appellate Body, 12 October 1998, paras. 79–97.

ones include the International Labour Organization's (ILO) Tripartite Declaration of Principles concerning Multinational Enterprises and Social Policy and the OECD Guidelines for Multinational Enterprises. The work on a UN Code of Conduct on Transnational Corporations, carried out by the United Nations Centre on Transnational Corporations, was discontinued in 1992 (Nollkämper 2006: 191ff.; Wallace 2002: 1077–96). The existing codes of conduct call upon TNEs to commit themselves to observe certain minimum standards regarding investment and other operations in developing countries, particularly in the areas of corporate conduct, technology transfer, and labour and environmental standards. Gross misconduct and the disregard of basic standards by apartheid South Africa drew the attention of the UNGA and UNGA resolutions repeatedly condemned Western companies for conducting business with the South African government and requested a cessation of business relations (UNGA Resolutions 39/72; 40/64 A; 41/35 A; 45/176 A). A recent effort to appeal to TNEs is called Global Compact, proposed by UN Secretary-General Annan (Annan 1999). The Global Compact Initiative calls upon TNEs to observe ten principles laying down standards on human rights, labour, environment, and anticorruption, derived from a number of comprehensive legal and non-legal instruments.

Apart from these rather cautious attempts at regulating TNEs, a few international treaties have been interpreted to recognize their role and accord them with a legal standing before courts or tribunals. Other treaties directly address TNEs, for example, granting them legal standing in investment disputes.[27]

While international law predominantly concerns interstate affairs, TNEs have taken measures to regulate their own affairs. First, they have established an internal normative order. For example, when the 1984 Bhopal disaster revealed shortcomings in corporate governance, TNEs in the chemicals industry responded to the severe repercussions the accident had on their public image; parent companies issued environmental standards and closely monitored their implementation by their subsidiaries.[28] Second, self-regulation is not limited to a TNE's internal affairs. States have created laws that pertain to transnational business. For example, the UN Convention on Contracts for the International Sale

---

[27] Cf. Art. 22 of the Convention on the Settlement of Investment Disputes Between States and Nationals of Other States.

[28] For an analysis of self-regulatory in the chemicals industry cf. Herberg 2005 and Herberg 2006.

of Goods (CISG) lays down rules for transnational sales contracts, for example the obligations of the buyer and seller. Yet, TNEs increasingly resort to a proprietary body of rules, termed *lex mercatoria*, an allusion to the medieval law merchant. Prominent examples of the *lex mercatoria* are the Incoterms, standard definitions applied in international trade and the dispute settlement system provided by the International Chamber of Commerce (International Chamber of Commerce 1999).

Governance capacity of civil society actors

Civil society actors play an active role in the international system, regardless of their status in international law. Although they do not replace the state, the dominant international actor, they show that governance is not exclusive to, or dependent on, the state. When humanitarian NGOs provide emergency help in the aftermath of natural disaster, they carry out tasks that are under normal conditions within the responsibility of states and the international community of states. In the absence of governmental regulations, TNEs operate within their own rules, and do not necessarily have to resort to the state-based legal system.

Governance without the state – through transnationally operating NGOs and TNEs – is a fact. However, what does this mean for governance by the state at the international level? It has been demonstrated that the state has laid the foundations for the purposeful and effective relocation of competences – governance capacities stemming from its sovereignty – to appropriate bodies. The question becomes of how governance by states relates to the efforts of civil society actors.

4
International Institutional Cooperation

In order to establish an effective system of public governance at the international level, states set up international organizations (IOs), endowed with competences that were previously inherent to the state. Consequently, IOs are "[t]he most obvious and typical vehicle for interstate cooperation" (Klabbers 2002: 28). With their creation and operation, states have institutionalized and solidified international cooperation.

The question is whether IOs remain an adequate vehicle for interstate cooperation. In the following section, the current academic literature on international institutional cooperation will be outlined. A brief historical account will explore the developmental steps of international institutional cooperation over the past 200 years. This review will serve as a basis for a description of the functions and performance of IOs today, pointing out their inherent deficiencies. This will eventually lead to an examination of alternative modes of transnational cooperation.

Creation of IOs

The principal objective of interstate relations was the preservation of peace, through either the maintenance of a system guaranteeing security and stability or the negotiation of peace treaties. Until the early twentieth century, peace and international security were mainly maintained through international conferences. These were convened on an ad hoc basis whenever issues arose that could not be resolved bilaterally through the usual diplomatic channels[29] The ad hoc character of

---

[29] Cf. Köck and Fischer 1997: 89ff. for an overview of the most important conferences.

such meetings and their unsystematic agenda prevented the erection of an institutional order for international policy issues; however, they occurred frequently enough that they were referred to as the *Concert of Europe*. Despite their exclusivity – only the major European powers were granted admission – a sense of community interest and awareness of interdependence began to evolve (Claude 1960: 21–2).

Certain successes notwithstanding, the conference system's ad hoc character and exclusivity were its most salient flaws. Procedural issues and the agenda had to be renegotiated for each conference, further delaying the process (Köck and Fischer 1997: 89ff.; Amerasinghe 2005: 1ff.). In many cases, conferences were convened on a post hoc basis, that is, they were usually convened in response to crises such as wars. Certainly the Vienna Final Act of 1815 tried to stabilize the power structure, aiming to safeguard future peace. However, the efficacy of such agreements was rarely assessed. The system's insufficiencies were often revealed with each new crisis. War would be the only indicator of failure of the system.

Interstate relations solidified first in apolitical areas. Growing economic interdependence and the inception of the Industrial Revolution at the end of the eighteenth century necessitated the coordination of technical regulations to facilitate trade, as bilateral relations alone could not satisfy the need for permanence and reliability (Delbrück 2001: 124–30; 2002b: 405). The Final Act of the Congress of Vienna is remarkable in this regard, recognizing in Arts 108–16 international watercourses and calling upon the abutting countries to take measures to facilitate navigation and trade on rivers. In an annex to the Final Act, special regulations on the navigation of the Rhine are laid down,[30] establishing the Central Commission for the Navigation on the Rhine, which still exists today.[31] Other river commissions followed, culminating with the principle of free navigation laid down in the Final Act (Amerasinghe 2005: 4; Seidl-Hohenveldern and Loibl 2000: marginal n. 208.9).

In other non-political, technical fields states began to create IOs – so-called AUs – that had the purpose of regulating technical matters. In addition to the growing codification of international law at this time,

---

[30] Annex XVI B, *Règlement particulier relatif à la navigation du Rhin*.

[31] The legal basis has been revised several times, cf. the Convention of Mainz (1831) concerning the navigation on the Rhine, the Convention of Mannheim (1868, revised Convention concerning navigation on the Rhine), and Strasbourg Convention (1963, amendment to the revised Convention concerning navigation on the Rhine).

the major structural change to international law during the nineteenth century was the emergence of institutionalized forms of international cooperation. By creating functionally limited organizations, states established a new type of international legal subject; however, states remained the masters (Delbrück 2001: 7).

Prominent examples of this development are the International Telecommunications Union[32] and the Universal Postal Union.[33] Although the notion of sovereignty was still equated with independence, states acknowledged the necessity of cooperation in transboundary affairs, creating appropriate institutions in this regard (Hobe 1999b: 259). AUs had a permanent character, with periodical conferences between state representatives and a secretariat to manage administrative tasks (Amerasinghe 2005: 4; Seidl-Hohenveldern and Loibl 2000: marginal ns 210–11). Their purpose was to function as clearing houses for information; only a few were granted the power to regulate certain matters independently (Claude 1960: 319).

An institutionalization of efforts to maintain peace and security has been envisaged by lawyers and philosophers for the past 800 years.[34] The idea was realized only in 1919 with the creation of the League of Nations (Köck and Fischer 1997: 101ff.). This event marked a new chapter in the process of the institutionalization of international relations. In comparison to the AUs of the nineteenth century, the League of Nations had an explicit political and universal character (Amerasinghe 2005: 5; Seidl-Hohenveldern and Loibl 2000: marginal n. 216). Another important feature, already visible in the AUs, was its openness. President Wilson, whose Fourteen Points gave the final impetus for the creation of the League of Nations after several government leaders in Europe and private groups advocated the creation of such an organization (Claude 1960: 37), envisaged it as a place of unlimited discussion (Wilson 1984). While participation in the conference was by invitation only, every state could obtain membership, thus making it a platform for smaller states and enforcing the principle of equality of states (Amerasinghe 2005: 6–7).

Although the League of Nations ultimately failed – mainly as a result of the unwillingness of states to implement its concepts – the

---

[32] Founded as the International Telegraph Union in Paris, 17 May 1865, renamed in 1934.

[33] Founded as the General Postal Union in Berne, 9 October 1874, renamed in 1878.

[34] Cf. Köck and Fischer 1997: 73–87 for an account of the history of ideas regarding IOs.

need for a universal international organization was recognized after World War II. With the main objective of preserving world peace, the United Nations Organization was founded in 1945, establishing a system of collective security (Seidl-Hohenveldern and Loibl 2000: marginal ns 220–1). With European states overburdened by war recovery efforts, close international cooperation became necessary, resulting in the establishment special organizations such as the Organisation for European Economic Co-Operation (OEEC), which was set up to implement the European Recovery Programme. The United Nations was also supported by a number of specialized organizations and a system of institutionalized economic cooperation, established in 1944 in Bretton Woods, with the creation of the International Monetary Fund (IMF), the International Bank for Reconstruction and Development, and the General Agreement on Tariffs and Trade (GATT) (Klabbers 2002: 21–2; Tomuschat 1999: 59ff.). Regional organizations also emerged after World War II, based on similar premises. This group includes the European Community (EC), which eventually resulted in the creation of a supranational order, whereby the organization effectively exercises supremacy over its member states (Seidl-Hohenveldern and Loibl 2000: marginal ns 222ff.).

Generally, states create IOs and endow them with the necessary authority to enhance the efficiency and effectiveness of their cooperation (Dicke 1994: 114ff.), that is, the costs of direct state interaction outweigh the costs of the IO (Abbott and Snidal 1998: 99). Consequently, states have created IOs to perform specific functions, which normally exceed the capacities of the individual state and require a specific structure. The spectrum of conceivable functions is very broad, ranging from the dissemination of information to the implementation of specific measures. Depending on the level of authority conveyed to a particular organization, it may even engage in the formulation of binding standards and laws.

IOs enjoy a certain degree of autonomy. This advantage provides them with the neutrality necessary to organize the settlement of international disputes. This distinguishes IOs from other international bodies. Their neutrality and relative autonomy translate into moral authority, which must be distinguished from their legal authority. An independent IO can promote intergovernmental activities and place items that might be otherwise be neglected on international agendas. Other than a powerful state, which will usually be suspected of bias towards a certain policy, IOs are usually considered to be neutral. Their neutrality also enables them to "launder" political concepts that might be unacceptable if advanced by one state. In this respect it is safe to say that "[s]tates establish IOs to act as a representative or embodiment of a community of states" (Abbott and Snidal 1998: 4–24).

The number of IOs has increased dramatically since World War II (Wessels 2000: 154ff.). As a result, the second half of the twentieth century can be considered the age of institutionalized interstate cooperation (Hobe 1998a: 528–9). Correspondingly, the number of international committees, boards, and panels has also increased over the years (Wessels 2000: 415ff.). Scholars observe an exponential growth of such committees, prompting the diagnosis of "committee-hypertrophy" (Id.: 428; Delbrück 1987: 398).

Deficiencies of IOs

IOs have always been flawed, their imperfections resulting from a poor institutional design, insufficient resources, and recalcitrant member states. These defects become particularly apparent in the age of globalization, where swift and concerted responses to problems are more important than ever.

The UN is a prominent example of an institution flawed by poor or outdated design. More than 50 years after their creation, the efficacy and efficiency of UN institutions are impaired by ingrained reflex responses, red tape, posturing instead of acting, and turf fights with one another because of overlapping or blurred competences (Junne 2001: 211–12). Presumably, these dysfunctional symptoms beset any bureaucracy (Barnett and Finnemore 2004: 8; 34ff.).

As a result of overlapping or unclear areas of responsibility, organizational problems arise that, especially in the field of international environmental policy, hamper effective international governance. The International Court of Justice (ICJ) considers the international institutional system of the UN to be coherently designed;[35] however, an examination of the competences regarding environmental affairs reveals a different picture, refuting the ICJ's assessment. A large number of IOs dedicate some effort to environmental issues. As a result of a mainly universal approach, IOs tend to cover environmental issues both horizontally – environmental media – and vertically – activities on all levels – so friction because of overlapping responsibilities and duplication of work is inevitable (Kilian 1987: 351). Overlapping competences, however, are a structural problem that IOs can address themselves. In

---

[35] ICJ, Advisory Opinion of 8 July 1996: 80, para. 26.

fact, cooperation between IOs has been established through several forums (Tietje 2002a: 54–5). Generally, these institutional shortcomings could be easily rectified through appropriate reforms. The ongoing reform process of the UN system exemplifies the tediousness of undertaking reorganization and adjustment to the challenges of today (Dicke 1994: 279–305). The problem is that global problems demand flexible responses and because of structural flaws, some IOs are unable to tackle complex global issues adequately.

This leads to the problem where one obstinate state can virtually paralyse an IO. Like any association, an IO can perform only as well as allowed by the appropriate resources its members are able and willing to contribute. If a key member disagrees with a certain policy, the IO may become ineffective. Compromises may be negotiated to formulate a policy that suits powerful and influential members and concessions granted to such members may ultimately impair the IO's credibility and undermine its moral authority (Barnett and Finnemore 2004: 169). Eventually, this damage will affect an organization's flexibility and effectiveness.

Globalization affects national administrations in such a way that regulatory matters transcend borders, falling only partially – if at all – within the jurisdiction of individual states and their agencies. As a result of their position in the international sphere, with their moral authority founded dually on their neutrality and autonomy, IOs are deemed competent to deal with all transnational or international matters, but this does not mean that IOs can legally assume responsibility for these areas. Typically, they lack the legal authority to regulate effectively the various transnational matters that cause global problems. IOs are created to perform specific functions and are endowed with the necessary authority to perform them effectively. Under the implied-powers doctrine, they may expand their competence, but while this doctrine may be invoked to secure the effectiveness of the organization, it may not lead to further restrictions on state sovereignty (Martinez 1996: 78–98). Thus, an IO must operate only within the authority conveyed by member states. Even where necessary, an IO may not generate its own competences.

IOs lack the necessary flexibility to carry out their mandate in the age of globalization effectively. They were designed at a time when the main objective was the preservation of peace – and not in the context of the numerous, highly fluctuating problems of the globalized world. This leaves the question of what institutional structures are instead required to tackle these pressing problems.

Transnational bureaucracy networks

In some respects, IOs epitomize an institutionalization of the idea of international cooperation for the twentieth century. In the face of globalization, the question becomes "what form international cooperation [will] take in the [21st] century?" (Raustiala 2003: 2). According to many scholars, empirical evidence indicates that transnational bureaucracy networks, which interlink national agencies, will take on this role.

By the 1970s, Keohane and Nye had already observed the emergence of transgovernmental relations, the "direct interactions among subunits of different governments that are not controlled or closely guided by the policies of the cabinets or chief executives of those governments" (Keohane and Nye 1974: 43). Officials began to coordinate their activities through informal communication channels, and engaged in coalition building with the goal of influencing domestic or international decision-making processes. Bypassing foreign offices, states ceased to act homogeneously in foreign politics. Additionally, IOs initiated transgovernmental relations, and their own staff were recruited for transgovernmental coalitions (Id.: 44–53). As regulatory matters become increasingly global, national regulatory agencies link with each other to form networks (Raustiala 2003: 3–4), a development that deserves closer scrutiny. The following section will briefly examine the nature of such networks and investigate both their benefits and flaws.

An introduction to networks

Policy networks have been in the research focus of political scientists for a long time. They are typically defined as "a set of relatively stable relationships which are of non-hierarchical and interdependent nature linking a variety of actors, who share common interests with regard to a policy and who exchange resources to pursue these shared interests acknowledging that cooperation is the best way to achieve common goals" (Börzel 1998: 254). Networks are an organizational form of governance (Id.: 255–9; Slaughter 2004: 40). They are a response to the growing interdependence of states and the requirement of cooperation. They are deemed to offer advantages to the two traditional forms of governance – hierarchy and market. Markets may fail and hierarchies may disregard minority rights (Börzel 1998: 260–1).

Building on the observations of Keohane and Nye on transgovernmental relations, it is only recently that international lawyers have started to focus on transnational bureaucracy networks.[36] In the mid-1990s, Chayes and Chayes pointed out that IOs and domestic agencies interact intensively; the organizations' secretariats often induce the creation of "transgovernmental elite networks" to facilitate the work and promote the organization's goals. Ultimately, transnational bureaucratic networks, coalitions of domestic authorities within the area of concern, emerge as a supportive structure (Chayes and Chayes 1995: 278ff.). Slaughter presented empirical evidence for transnational networks linking government agencies, and put this observation into the broader context of global governance (Slaughter 1997).

On the basis of available data, networks involving government actors can be distinguished on the basis of their setting (Slaughter 2004: 45ff.). The first group concerns networks within IOs, the second concerns those within the framework of executive agreements, and the last group concerns spontaneous networks. Examples of networks involving IOs are trade ministers' meetings within the GATT framework, and the convening of defence and foreign ministers for NATO. Actually, the close connection of these meetings to the organizational setting in which they take place makes it hard to distinguish them from actual IO organs. The second type of network is based on the structure of the treaty itself, separate from its formal institutions. Because these structures are in place for the implementation of the treaty, these networks can also be termed "implementation networks," since their objective is to support the realization of the treaty's goals. One example of an implementation network arises with regard to the Convention on Biological Diversity (1992), where certain government agencies have formed a network working towards the further development of the convention (Korn 2004). The last type of network exists outside of IOs, emerging spontaneously; here, government agencies band together to form a loose institution for regulatory cooperation.

A detailed study recently explored the workings of international financial regulatory organizations (IFROs), such as the Basle Committee on Banking Supervision, the International Organization of Securities Commissions (IOSCO), and the International Association of Insurance Supervisors (IAIS) (Zaring 1998).[37] Reference to this study helps to illustrate the phenomenon of transnational bureaucracy networks. IFROs were created by government agencies – central banks, securities

---

[36] Slaughter 1997; 2000a; 2004; Zaring 1998; Picciotto 1996; 1997.

[37] For an investigation of these entities from the German perspective cf. Möllers 2005: 355–61.

commissions, or insurance regulators – not by the state itself. Their structure is informal, and is not based on an international treaty: instead, agreements between agencies form the basis of IFROs. In the cases of IAIS and IOSCO, these agreements were cast into a legal form: IAIS is incorporated as a not-for-profit organization in Illinois, and IOSCO derives its legal personality from an act of the Quebec National Assembly. They derive much of their flexibility through their informal status, demonstrated by their minimal set of internal rules, and lack of transparency – these organizations maintain a very low profile. They do not have the capacity to issue legally binding orders, but agreements reached within IFROs can influence national lawmakers (Zaring 1998: 301–4).

Transnational bureaucracy networks perform a variety of functions: collecting and disseminating information, enabling coordination and cooperation in regulatory matters, and fostering the harmonization of rules and standards (Slaughter 2004: 131). This last aspect is probably the most interesting. Transnational bureaucracy networks may reach a consensus on a certain issue and develope codes, principles, or recommendations on this basis. Because of their advocative character, and bolstered by the authority of the participating actors, these instruments are a specific type of law (Id.: 177ff.).

Transnational bureaucracy networks are characteristically non-hierarchical, informal structures, interlinking national agencies in a specific policy area with the aim of addressing common problems through exchanging information, coordinating strategies for action, and formulating common rules. IOs play a specific role in the creation and maintenance of networks; like a "spider's web," networks take advantage of the existence of IOs, using them as points of attachment (Picciotto 1996: 1020–39).

The emergence of such networks is attributable to two characteristics of the modern state: its openness on the international sphere and the disaggregation of its administrative structure. Only a constitutional cooperative state would allow its agencies to operate beyond its national borders. Furthermore, the breaking up of state administrations provides agencies with the degree of autonomy necessary to connect with their foreign counterparts and participate in such networks.

Advantages and deficiencies

Networks, in comparison with IOs, have a number of advantages. It has been shown that IOs can be cumbersome and thus ineffective, as a result in part of their rigid procedural rules, dwindling resources, and lack of state support. Networks, on the other hand, are endowed with a high degree of flexibility, stemming from their informal and non-hierarchical nature (Raustiala 2003: 24; Slaughter 1997: 193). New challenges will not be met by a rigid structure: networks are pliant enough to make any adjustments necessary to address new issues and this also allows them to find ways to collaborate with non-state actors, benefiting from their expertise and resources (Slaughter 1997: 195). Furthermore, networks also assist in speeding up the global problem-solving process. The international system traditionally regards the state as a unitary entity. Once a domestic position is agreed upon, it is presented to the international community. Negotiators seek to reach a position supported by international consensus, this must then be implemented at the national level. Instead of this time-consuming bottom-top-bottom approach, networks allow competent government officials to become directly involved in the problem solving process from the outset (Slaughter 2004: 170).

Because of their lack of formality, transnational bureaucracy networks largely rely on "soft power" (Nye 2004: 5–32; 99–125), that is, persuasion, flattery, expertise, and peer pressure, in their decision-making processes (Raustiala 2003: 24). This does not distinguish them from IOs; however, while IOs operate entirely on the international level, national agencies operating in transnational bureaucracy networks can resort to state authority – hard power – to implement their decisions once a course of action has been agreed (Slaughter 2004: 167).

Transnational bureaucracy networks offer two advantages to states. First, because they are established within existing channels, they neither require the establishment of a new bureaucracy, nor create new organizational interests (Junne 2001: 219). Second, networks conserve state sovereignty. Formally, the state retains its sovereign rights and does not have to relinquish power to another organization. Authorities participating in the network operate within state boundaries and the state maintains its monopoly on the legitimate use of force (Slaughter 2000b: 201).

In sum, transnational bureaucracy networks offer the advantage of expediting the global decision-making process, through taking a problem-oriented approach to issues and implementing solutions much more easily.

Transnational bureaucracy networks are not without flaws. The main concerns are the control – or lack of it – of their activities and the legitimacy of their authority (Slaughter 2000b: 203). These actors are unelected, yet they engage in global regulation. Moreover, their functional approach to problem solving could dissociate them from the social, economic, or political concerns of citizens, turning them into a massive, globe-spanning technocracy (Slaughter 2004: 219).

Their informal character, which contributes greatly to their flexibility, proves to be problematic. Generally, informal procedures and committees are not new phenomena in administrative law. In fact, informal mechanisms, for example gentlemen's agreements or consultations, used to settle conflicts between a regulatory agency and citizens, are employed far more often than formal mechanisms like administrative acts or rules (Bohne 1981: 74ff.).

However, the informal approach, combined with the transnational aspect of their activities, could render such networks virtually invisible, allowing them to operate below the radar of those who would hold them to account. Such networks often rely on gentlemen's agreements, or slightly more formalized MoUs; however, these agreements are rarely published, leaving the public is unaware of their existence (Picciotto 1996: 1047; Raustiala 2003: 49). The informal character of these agencies also raises doubts whether it is possible for ministries or parliament to supervise them adequately. Thus, their activities run the risk of slowly eroding political and administrative hierarchies and may even undermine parliamentary oversight. Empirical evidence suggests that agency officials still operate in transboundary contexts under the shadow of political and administrative hierarchies (Wessels 2000: 429ff.), but then one has to wonder whether the shadow grows larger or shrinks.

Contributing to the legitimacy problem is the exclusive character of some networks, which further masks the decision-making process (Benz 1995: 202–3). In certain contexts – particularly antitrust or tax regulation – informality and confidentiality are essential to the functioning of the network (Picciotto 1996: 1049). Nevertheless, the system of checks and balances inherent to the modern state requires the government to control the administration. Controlling such obscure networks may become difficult.

In sum, transnational bureaucracy networks have been diagnosed with a "chronic lack of legitimacy," the main concern being whether their decision- or rulemaking processes satisfy the requirements of democratic legitimacy (Id.: 1047).

Despite these deficiencies, scholars view transnational bureaucracy networks as an effective method of global governance, preferable to a "supranational bureaucracy, answerable to no one" (Slaughter 1997: 184ff.). Some suggest that these networks might supersede the traditional mode of international cooperation achieved by means of international treaties and IOs (Raustiala 2003: 71). Networks, they argue, represent the "real new world order" (Slaughter 1997; 2004). However, the research carried out up to now points towards a strong role for IOs in the emergence and maintenance of networks. Notwithstanding the question of whether this assessment reflects adequately what is actually happening in the inter- and transnational arena, the democratic legitimacy of such networks remains a major predicament that is still to be resolved. Transnational bureaucracy networks certainly have benefits, linked to their unique composition. Therefore, legitimizing their activi ties might not require rigid and comprehensive supervision, but rather only an occasional correcting intervention or complementary participation (Scharpf 1991: 631–2).

5
Law and Globalization

Law is probably the most important and powerful governance tool at the state's disposal. Backed up by the monopoly of force, the state can institute legislation to influence the behaviour of the natural and legal persons within its territory. For matters beyond their territorial boundaries, states conclude bilateral or multilateral agreements, erecting an international legal framework. With the guarantee of state intervention to resolve conflicts and uphold the legal order, law has become a reliable fabric underlying society.

In modern times, the stability and adequacy of national and international law have both been questioned. First, the state faces the challenge of how to use law to address risks arising from industrialization. Static legal rules have proved inept at controlling technology; consequently, law is considered to be "antiquated" (Wolf 1987: 357). Law faces the challenge of dealing with uncertainty in many areas. Examples are pharmaceuticals, nanotechnology, genetics, and chemicals. Each area holds its own risks. The likelihood of the occurrence of negative effects can be estimated, but essentially, law has to deal with uncertainty. Lawmakers are overburdened with the elaboration of rules addressing such risks. In this task, they often resort to indefinite legal conceptions like "best available technique." Physicists and engineers, for example, will have to determine what kind of technology currently represents the state of art (Id.: 365ff.). Thereby, the governance of key aspects crucial to the state's task of protecting its citizens from harm is left not to elected officials, but to technicians and engineers.

Today, law is faced not only with uncertainty but also with the fact that the items to be regulated have become transnational – thus frequently evading its grasp. This warrants a close examination of the state of law in the age of globalization. First, national law has a certain degree of flexibility, but its validity ends at the borders. International law may have greater reach, but it is rather rigid and applies only to states, and not to transnational actors. Furthermore, the private governance measures of civil society actors result in a quasi-legalislative order, with an unclear relation to the state-based orders of national and international law. This raises the question of whether national and international law systems are adequate governance resources in the age of globalization.

The following section will investigate the current state of law, examining law and legal theory in the age of globalization with the aim of developing a working definition of law for the remainder of this book. It will also be necessary to look at the adequacy of, and changes to, international law as well as the emergence of other forms of law in order to gain a full picture of law as a source of global governance.

International law

International law in its current form is considered to be incapable of providing the flexibility necessary to cope with the dynamics of globalization, particularly in regards to the protection of the world environment. Traditional sources of international law are not supposed to have the necessary potential for innovation (Schreuer 1983: 243).

The lawmaking process is both slow and costly. Several stumbling blocks hinder the multistage drafting process. Consensual decision-making, a corollary of sovereignty, impedes the conclusion of effective agreements as the slowest party determines the pace of the drafting process and the lowest common denominator is the outcome of negotiations (van der Lugt 2002: 226). Even ineffectual treaties run the risk of being rejected by national parliaments during the ratification process. Once a treaty as been ratified, some states simply lack the resources to implement their obligations effectively, in spite of their willingness to comply with treaty requirements in principle (Reinicke and Witte 2000: 88ff.; Neuhold 2005: 40–3; Sand 1992: 240). Another shortcoming of treaties is their inflexibility (Neuhold 2005: 16 7). Procedures designed to adapt treaties to meet current challenges tend to be as cumbersome as the initial drafting.

Some scholars even contest the legitimacy of international law, considering it an outdated form that will be imminently replaced by new forms of law stemming from an emerging global society.[38] According to their assessment, normative rules instituted by civil society actors have

---

[38] Cf. Zumbansen 2000, who speaks of the future past of international law.

become a substitute for the international legal order. With the waning powers of the individual state, the necessity for an international law supposedly disappears.

There have, however, been developments in international law that ensure its flexibility, and thus effectiveness with regards to solving global problems. A key principle of international law is *pacta tertiis nec nocent nec prosunt*, according to which international treaties bind only the contracting parties: third parties are not bound by such agreements. A treaty may regulate matters of global importance, but if a state views the deal as being not in its best interest and abstains from ratifying it, the state will remain unbound. A particularly dramatic example concerns unsustainable resources. Fisheries regimes may attempt to protect and sustain stocks. However, if those states with large fishing fleets do not join the regime, it is practically useless as a means of achieving the sustainability of stocks (O'Connell 1993: 303). Similarly, the persistent objectors are generally not bound by emerging customary international law. The so-called *obligations erga omnes* (in relation to everyone) principle – a concept closely related to *jus cogens* (peremtory norm) laid down in Art. 53 of the Vienna Convention on the Law of Treaties (VCLT) – constitutes an exception to this rule (Tomuschat 1999: 81–4; Kadelbach 1992: 32–3, 178). An *obiter dictum* (incidental remark) of the ICJ ruling in the Barcelona Traction case suggested that norms *erga omnes* could be applied to bind not only specific states, but the international community as a whole;[39] some treaties create an objective, comprehensive regime, and therefore, must be ultimately respected by all states (Delbrück 2002b: 415–16). Such rules can be found, for example, in international fisheries law and are of special significance in international criminal law (Hobe 1999b: 275; Brownlie 2003: 568). The enforcement of these rules, however, is unclear (Hobe 1999b: 275; Zemanek 1998: 856). The increasing acceptance of norms *erga omnes* indicates the transformation of international law from a legal order based on expressed or implicit consent to an objective legal order (Delbrück 2002b: 417). The fundamental element of international law has moved away from the will of states to a system of common values, as evidenced by the recognition of human rights and the abolition of slavery in the Final Act of the Congress of Vienna (Frowein 2000: 428–31).

This change raises the issue of the legitimacy of such norms, since the particular group of states involved in setting such norms acts as

---

[39] ICJ, Judgement of 5 February 1970: 32.

a global legislator, claiming authority over others, and bypassing the principle of the equality of states. Delbrück notes that norms *erga omnes* will likely be confined to those rules that touch upon the "international public interest or international community interest" (Delbrück 2002b: 418). The principle of *pacta tertiis nec nocent nec prosunt* would remain applicable in all other cases. NGOs also influence and contribute to the lawmaking process, enhancing the legitimacy of norms stemming from *erga omnes*. However, the role of NGOs and other non-state actors as a legitimizing factor is contested, since they represent a limited constituency.

As a response to the cumbersome international treaty-making process set out in Art. 39 (f) of the VCLT, various new structures have emerged to make treaty regimes more flexible.[40] To ensure a high degree of flexibility, framework convention is increasingly used as a model. This approach is often supplemented by simplified amendment procedures (Tietje 1999: 36–9). Framework conventions provide an institutional basis for further political and scientific cooperation and incremental regulatory measures (Ott 1998: 269). Technical details are generally regulated in annexes, which can easily be amended by majority votes of the parties on the basis of new scientific findings or a performance review. Since full agreement is not a requirement, states do not have to settle for the lowest common denominator (Bleckmann 1995: 751; Sand 1992: 254–6); those states that reject the negotiated changes can opt out so that they are not legally bound by amendments. Instead of the implementation of the treaty deemed acceptable by a majority of states being stalled because of the objections of a few states, the treaty regime remains in operation and enforceable. The framework convention itself does not contain obligations; its success depends on the adoption of amendments (Beyerlin 2000: 42–3). Accordingly, framework conventions cannot fulfil their purpose if the parties to the convention are not willing to adopt amendments or if important parties choose to opt out. The decision of the USA to reject the 1992 Kyoto Protocol to the United Nations Framework Convention on Climate Change is an example for this problem.

Another means of enhancing the flexibility, and thus efficacy, of treaties is to empower the treaty secretariat with the authority to interpret unclear provisions (Tietje 1999: 39). For example, Art. XXIX of the Articles of Agreement of the International Monetary Fund tasks the Executive Board and the Board of Governors, respectively, of the

---

[40] For an overview cf. Hingst 2001: 163–75.

International Monetary Fund (IMF) with the authoritative interpretation of contested provisions of the IMF agreement.

Decisions originating from competent private or public bodies may also be legally incorporated into the treaty regime (Id.: 40). Evidence of this practice can be found in Arts 2.4 of the Technical Barriers of Trade (TBT) Agreement and 3.2 of the Sanitary and Phytosanitary Measures (SPS) Agreement, which refer to "international standards" as a way of rulemaking efforts outside the World Trade Organization (WTO) context.

New instruments were also introduced in international environmental law to enhance treaty performance. For example, the 1987 Montreal Protocol to the 1985 Vienna Convention for the Protection of the Ozone Layer introduces measures to enhance cost-efficiency and the promotion of innovation. The United Nations Framework Convention on Climate Change and its Kyoto Protocol institute an economic system, providing for the trading of emission rights (Schuppert 1998: 1ff.).

Circumstances at the beginning of the twenty-first century were quite different than during the 1950s when international law last underwent fundamental changes. There have also been several recent changes: apart from the growing interdependence of states, IOs have gained importance as international actors representing common public interests and aiming at the preservation of common global goods. Non-state actors, such as multinational enterprises or NGOs, have also gained increasing relevance as spokespersons for particular civil society interests. International law started to transform into an "internal law" of a world community, extending beyond states and IOs to NGOs, TNEs, and individuals (Delbrück 1993: 725; 2002: 401–2). Furthermore, the issues that the international community must deal with have changed and become more complicated. While the prevention of war remains an important goal, other major challenges have also arisen. Infectious diseases can spread throughout the world within days because of the high mobility of individuals. Financial and economic stability is crucial because of the highly integrated nature of the economic system. Global environmental problems affect everyone, endangering the planet's ability to sustain human life. The question now is of whether the increasing interdependence resulting from globalization will lead to further structural changes to international law.[41]

---

[41] Hobe 2002: 385 considers a paradigm change towards an international law on globalization.

Soft law

International lawyers usually refer to Art. 38 (1) (a)-(c) of the ICJ Statute to define international law. However, many scholars point out that this particular provision does not constitute a *numerus clausus* of modes to create international law. In fact, the categorization describes only the external forms traditionally employed by states to express their legal will, but does not prevent them from establishing new sources (Verdross and Simma 1984: § 518; Tomuschat 1972: 111–12). The article is thought to primarily determine the scope of jurisdiction of the ICJ. Furthermore, as it is part of an international treaty, it may be subject to change, depending on the will of the parties to the statute. It serves as an indicator for international law; however states employ other, diverse, measures that they consider adequate for resolving specific issues or problems (Tietje 2003b: 30–1).

Throughout recent decades, instruments have been developed and employed which do not fit into the triad of sources laid set out in Art. 38 (1) (a)-(c) of the ICJ Statute. International lawyers have referred to such instruments as "soft law," because although the contracting parties lack the will to be legally bound by an agreement, they conclude agreements resembling legally binding accords in many ways. In this sense, the term soft law appears paradoxical (Dupuy 1991: 420).

Taking a binary view, more conservative international scholars consider that law is obligatory (i.e. hard) or it is not law at all (Thürer 2000: 456). Some argue, therefore, that soft law is redundant or even undesirable, as it undermines the "blissful simplicity" of law and does not contribute to the solution of political problems or a pathological phenomenon, thus blurring the line of what is normative (Klabbers 1996; 1998: 387–91; Weil 1983: 415ff.).

The above position ignores two important aspects. First of all, international treaties can also be soft instruments. While they might, in a formal sense, fulfil the requirement necessary to be considered international law, that is, in the sense of Art. 38 para. 1 (a) of the ICJ Statute, they may also contain provisions phrased too vaguely actually to impose an obligation (Dupuy 1991: 429–30). Agreements can be significantly weakened either because they are imprecise or unclear or because they delegate to third parties the authority to implement, interpret, and apply the rules or resolve conflicts (Abbott and Snidal 2000: 422).

These drawbacks notwithstanding, blissful simplicity is certainly not an end in itself. Strict dismissal of soft law ignores its legal and practical relevance. Many states incorporate nonbinding arrangements into their domestic legal order (Kunig 1989: 534). The constitutions of several newly independent countries have modeled their catalogue of human rights after the Universal Declaration of Human Rights (UNGA Res. 217 (III)), which in point of law is nothing more than a UNGA resolution (Schreuer 1983: 249). Similarly, the CSCE Final Act triggered constitutional revisions and legislative measures (Schreuer 1983: 249–50). Domestic courts may even refer to soft law that has not yet been transformed, in order to determine an infringement of the international *ordre public* (Geiger 2002: 191; Kunig 1989: 535). Courts in the Netherlands, Belgium, and Germany have repeatedly employed the Universal Declaration of Human Rights in this regard (Schreuer 1983: 257). It should be noted, however, that domestic courts – at least in Germany – were very reluctant when confronted with soft law arrangements. Therefore, soft law must be regarded as a broader, less formal kind of international law. Its role in international governance has to be examined.

International lawyers broadly distinguish two types of soft law (Thürer 2000: 454ff.; Hobe and Kimminich 2004: 198). The first is made up of IO resolutions, which are nonbinding in most cases (Seidl-Hohenveldern and Loibl 2000: marginal n. 1547ff.). This so-called secondary law flows out of the original treaty establishing the IO (Id.: marginal n. 1502ff.). An important example of this kind of soft law is the Universal Declaration of Human Rights. States may feel obliged to implement the Declaration of Human Rights for moral or political reasons. From a legal perspective, it has been adopted as a UNGA resolution, and is thus nonbinding. However, this does not immediately render such commitments meaningless from a legal perspective. In its ruling in the Nicaragua case, the ICJ recognized the legal value of UNGA resolutions. According to the court, these often play an important role in the creation of international customary law.[42]

Agreements between states, usually concluded at interstate conferences, are the second type of soft law. Prominent examples include the Final Act of the CSCE and Agenda 21, concluded at the 1992 United Nations Conference on Environment and Development (UNCED). States approach the creation of such agreements with great care, as if negotiating an international treaty (Dupuy 1991: 429). Often these norms serve their own distinct purposes, and thus, cannot be considered simply to

---

[42] ICJ, Judgement, 27 June 1986: 97–109; ICJ, Advisory Opinion, 8 July 1996: 254–5.

have an auxiliary character as a subsidiary source in the sense of Art. 38 (1) (d) of the ICJ Statute (Riedel 1991: 63). One such instrument is the MoU, which has a non-legal character reflecting the contracting parties' will not to be legally bound to implement the provisions. This is revealed in the chosen terminology through explicit provisions pertaining to the status of the agreement or refraining from registering the agreement in accordance with Art. 102 of the UN Charter (Aust 2000: 27ff.).

Soft law is clearly more than mere words or political posturing. In fact, it is identified as one possible instrument to meet the demands of globalization, where "hard" law has proven inadequate (O'Connell 2000: 102). As a result of its flexibility, variability, and non-obligatory character, which allow experimental solutions, it is considered an adequate global governance tool (Id.: 113; Neuhold 2005: 47ff.).

Soft law's flexibility makes it a useful instrument for international governance, as this allows faster responses to new demands. Low contracting and sovereignty costs are key factors for promoting flexibility. Contracting costs are lower because drafting, concluding, and amending such instruments can be carried out much faster than for a legally binding agreement (Abbott and Snidal 2000: 434ff.). Treaties legally bind states, which may fear losing authority over decision-making processes when treaties include provisions delegating authority to another entity and may consider such scenarios to impinge on their sovereignty. Nonbinding instruments, but also ambiguous provisions or the omission of the delegation of authority, can motivate reluctant states to enter into such agreements (Id.: 436ff.). Another aspect of soft law agreements that enhances their flexibility is the fact that their nonobligatory character allows parties to resort to experimental approaches toward problem solving much more easily than in cases of legally binding international treaties (O'Connell 2000: 109–10). After initially assessing the suitability of an instrument, states are free to abandon the implementation of the arrangement if it does not yield the desired results (Abbott and Snidal 2000: 442). Hence, soft law appears to be especially useful for the regulation of highly complicated technical matters. Finally, soft law instruments are open to anyone and can therefore incorporate civil society actors. While international law only barely takes notice of NGOs or TNEs, these groups can easily be included in the creation and implementation of soft law instruments.

Despite the benefits of soft law, the overall goal of the international community remains the creation of hard international law with obligatory character and enforceability as important main advantages

(O'Connell 2000: 109–12). It is thus possible for soft law to have a preparatory character. If the nonbinding rules are proven to work effectively, they may ultimately be recast as "hard" law (Thürer 2000: 458). Soft law thus lays a foundation for creating legally binding instruments, by either launching or catalyzing the drafting of international treaties or by pointing to the formation of international customary law (Beyerlin 2000: marginal n. 141).

However, soft law also has disadvantages and can pose dangers. Unlike treaties that legally bind a state, soft law instruments such as MoUs do not have to undergo constitutional procedures and are confidential insofar as their publication is not usually required, as is normally the case (Aust 2000: 35ff.). Thus, parliamentary or other forms of democratic control may be bypassed, to the effect that the public is not aware of interstate arrangements and the conduct of affairs (Thürer 2000: 458). Furthermore, soft law can become legally relevant. It can become a criterion for the determination of good faith, resulting in the evocation of estoppel by one party (Id.: 457; Aust 2000: p. 45). Here a danger might be that a soft law instrument gains more significance than intended by its creators.

Transnational law

During the 1950s, Jessup recognized that the term "international law" was too narrow to cover all of the rules relating to transboundary affairs. As a result, he coined the term "transnational law" to describe "all law which regulates actions and events that transcend national frontiers. Both public and private international law are included, as are all rules which do not wholly fit into such standard categories" (Jessup 1956: 2; 106). In fact, this perspective on transnational law shatters the strict dichotomy of municipal and international law, recognizing the interplay between these legal orders in matters that transcend national jurisdictions.

Nowadays, the term "transnational law" is used frequently by scholars of international law or international relations, and thus requires clarification. Friedman, for example, uses the term to describe "norms and institutions which span, are valid in, or apply to more than one country or jurisdiction," whereas a regime is only transnational "if it has the force of law, or the force of force behind it" (Friedman 1996: 66). However, this suggests a hierarchy of norms, whereas a norm system encompasses several countries or jurisdictions and enjoys priority over the covered domestic legal orders. A legal order with these features is usually – and aptly – defined as supranational, with emphasis on the prefix. European law directly affects the citizens and life in the member states[43] and is superior to the national legal orders, making it the proto-typical example of a supranational legal system.

Building on empirical surveys of the legal systems regulating inter-national economic affairs, others have expanded on Jessup's approach. While Jessup aimed to overcome the artificial separation of national and international law, authors today apply the term "transnational law" to the autonomous norm systems set up by civil society actors. Hence, transnational law is understood as an autonomous legal order sepa-rate from national or international legal orders (Calliess 2002: 186ff.). However, this does not mean that it is sealed off from the national and international legal orders. The case of *lex mercatoria* is used to illustrate how transnational economic law is "transnational in the sense of an interlocking plurality of various subjects of law, sources of law, and cor-respondingly levels of law-making" (Tietje 2002b: 407).[44] Thus, trans-national law is connected with legal orders of diverse provenance, the national legal order being but one of them. Nevertheless, it still main-tains its relevance as the place where the norms become obligatory and can be enforced (Id.: 416–17).

Transnational law has repercussions for legal theory. It touches upon the core questions of jurisprudence: what is law and who makes it? The traditional approach to answering these questions begins with the state and the concept of legal positivism. Hobbes already considered laws to be commands of the sovereign (*auctoritas non veritas facit legem*) (Hobbes: 1991: 136ff.). In the nineteenth century, Austin further devel-oped the concept of legal positivism. Laws are commands, set by the sovereign, noncompliance with which entails sanctions (Austin 1885: 88–9). Kelsen further elaborated this concept of law, pointing out that coercion is the determining feature of law, setting it apart from reli-gion and morals (the separation thesis). The threat of coercive measures brings about the desired social conduct (Kelsen 1967: 4). Thus, law is a normative coercive order (*normative Zwangsordnung*), which is founded on a basic norm (*Grundnorm*) (Kelsen 1960: 45ff.; 196ff.). The legal posi-tivist approach to law is widely abundant in today's legal studies. Hart declares that law is "...what the Queen in Parliament enacts..." (Hart

---

[43] Cf. Art. 249 of the EC Treaty and the ECJ jurisdiction on the direct effect of Directives. Cf. also the discussion of European law and sovereignty *supra* 17f.

[44] Translation by the author.

1994: 107). According to Black, "law is governmental social control … the normative life of a state and its citizens, such as legislation, litigation, and adjudication" (Black 1976: 2). Similarly, Dreier contends that "law is the entirety of norms, which belong to the constitution of a state based or interstate based norm system …" (Dreier 1986: 896).[45] The state is the sole source of law, having the monopoly legitimately to create law and enforce it. From a positivist perspective, state-based law gains its standing from the state's monopoly of force and thus is set apart from other rules such as customs, morals, or private rulemaking.

Starting from this assumption, some scholars contest the quality of international law based on the notion that no supreme authority exists to enforce it and that a basic cannot be identified (Hart 1994: 2327). Hoebel calls it "primitive law" for this reason (Hoebel 1968: 418). Measures exist that can certainly be deemed as sanctions for non-compliant behaviour (Kelsen 1967: 16–173), for example, retorsion or reprisal as responses to unfriendly acts, measures in accordance with Chapter VII of the UN Charter or compensation and suspensions of concessions in accordance with Art. 22 of the Understanding on Rules and Procedures Governing the Settlement of Disputes (DSU). However, while state law is enforced by a superior power, the principle of state equality prevents the existence of such a power at the international level. Force, however, is not the only mechanism to ensure legal compliance (Berber 1975: 15–16). As there is no international legislature functioning on a level comparable to domestic parliaments and no form of comprehensive and compulsory jurisdiction exists, treaty compliance can be accomplished through "managerial" means, that is, through the application of soft power via persuasion, cooperation, and an orientation toward problem solving as opposed to "coercive" means (Chayes and Chayes 1995: 3).

Obviously, strict notions of law do not and cannot apply in intergovernmental affairs. Hence, international law could cautiously be defined as "the normative expression of the international polity having States as its basic constituent entity" (Henkin 1989: 21).

Legal positivism has been criticized extensively,[46] mainly on the basis of the separation of law from morals, ethics, or customs. The various points of criticism will not be repeated here, but it is noted that because of its state-centredness the positivist approach is insufficient for gaining a clear picture of legal interactions in the age of globalization.

---

[45] Translation by the author.
[46] Cf. Alexy 2002: 39ff. for an overview of the discussion.

States are the dominant actors in the international system, but they are not alone. Consequently, the international, state-based legal order is not necessarily the only conceivable normative system, as it excludes TNEs and NGOs. The normative efforts of non-state actors like TNEs and NGOs, or those of informal transnational bureaucracy networks, and the ramifications of their operations cannot be revealed and assessed through filters that define law as state-centred.

Much more promising in this regard is a sociological approach that is not tied in with a *Grundnorm*, but explains law from its reference to society (Luhmann 1983: 23). Luhmann assumed this position and conceived of law functionally as opposed to the formal approach of legal positivism. In his view, the world is a complex place with many possibilities for action, from among which one must be chosen. Outcomes of these actions are not easily predicted. Thus, the world is also a contingent place where expectations can be disappointed (Id.: 31). Law is a social structure designed to reduce complexity and stabilize expectations. Accordingly, law is not so much a coercive order, but one that facilitates and stabilizes expectations (Id.: 100; 115). This functional perspective recognizes all those norms as law that can in fact substitute for state-based law (Röhl and Magen 1996: 20). In other words, any norm that is effective in the sense that its addressees adhere to it and that may be enforced in any way – including by cautious means – must be considered law.

This book will employ a broad definition of "law," based on the sociological approach outlined above: all rules – phrases with a prescriptive content (as opposed to a descriptive content) – that are backed up by some kind of authority are law.

The approach of legal pluralism recognizes the legal character of norms which do not have their origin in the state and are not backed up by its authority (Lampe 1995: 8). Recognizing law as social rules other than those backed up by the state does not negate the idea that law, in principle, needs to be supported by some sort of authority (Pospíšil 1982: 136). In this context authority means the power to order the application of the rules. Of course, the aspect of authority also entails the problem of legitimacy. This definition leaves the legitimacy of authority – an important question relating to the reason why rules should be accepted – untouched.

This approach towards law calls into question the de facto state monopoly on lawmaking. Monopolization of lawmaking as practiced in the modern state is a relatively recent development that has become a key feature of modern statehood; the law of non-state actors such as churches or sports associations exists only by grace of the state (Reinhard 2000: 281ff.).

Yet, within one state several quasi-legal orders coexist alongside the state-based legal order. The state does not hold a comprehensive monopoly on lawmaking. Its monopoly to apply force legitimately is limited to physical force, so other actors may set rules and enforce them by exerting social or economic force (Kirchhof 1987: 107–38).

This phenomenon, "a situation in which more than one body of laws or set of norms exist inside a single legal jurisdiction, country or other entity" (Friedman 1996: 67; Lampe 1995: 8) is usually described as legal pluralism. It is a common occurrence in colonized countries, where the state-based legal order exists coequally and overlaps with indigenous religious and customary laws. However, legal pluralism is not exclusive to such countries; the coexistence of legal orders can also occur in modern societies, depending on the concept of law (Merry 1988: 869–70). Ehrlich observed in the early twentieth century that within the modern state homogeneous, positive law coexists alongside "customs" or other legal orders. He noticed that societal norms adapt more quickly to changes and are highly significant to their addressees, in fact much more so than state-based norms. These constituted the "living law" not laid down in legal provisions, but nevertheless governing virtually any conceivable aspect of life (Ehrlich 1989).

In the 1970s, the positivist model of law was again contested by the work of Pospíšil, who showed that both modern and ancient societies are governed not only by positivist law ordained by the state, but by a plurality of coexisting and sometimes overlapping legal orders of various origins (Pospíšil 1982: 136–71).

Legal pluralism is not limited to the national sphere. *Lex mercatoria* is the most salient example of this development, receiving the most attention in legal and sociological literature (Cf. Teubner 1997: 8ff.; Robé 1997: 50ff.). In the absence of a state-based legal order for transnational economic activities, TNEs have created their own rules pertaining to the conclusion of contracts and the settling of disputes.

The extensive set of rules governing Olympic sports (Adolphsen 2002), international construction law (Molineaux 1997), and the aforementioned *lex mercatoria* indicate that civil society actors of diverse provenance create normative orders on the global level for a variety of purposes. On the national level, the state no longer holds a monopoly on lawmaking (Gessner 2002: 297ff.; Snyder 2000: 105ff.). Globalization thus gives rise to a global law which is not arranged in a coherent and hierarchical order, but instead made up of a multitude of normative systems, many of which originate not from states, but rather from private actors (Teubner, 1997: 4–5; Günther 2001: 539ff.). This legal pluralism not only acknowledges the coexistence of legal orders – a phenomenon that is hardly alien to practitioners or scholars of international or European law – but also suggests that these legal orders may have their origin in entities other than the monolithic state (Anders 2004: 39–40). With a broader definition of law, one can explore the various normative activities in their social fields and trace the interactions of and overlaps between the legal orders (Id.: 51).

The consequences of global legal pluralism for the state and law are assessed differently. One position, mainly advanced by Teubner, observes the shift of lawmaking away from the state, overburdened with the effects of globalization, to civil society actors. Consequently, global law emerges "from the social peripheries, not from the political centres of nation states and international institutions" (Teubner 1997: 7; 13). The plurality of global law systems is transnational, limited not by territory, but rather separated internally according to regulatory sectors. These "private regimes" were framed by the state-based legal order on the domestic level, but their transnational character allows them to evade the grasp of state law. Ultimately, globalization brings about a loss of significance for traditional lawmaking processes and an end to the coherence of law (Id.: 8; Teubner 2000: 439–40; Teubner and Fischer-Lescano 2004: 1000ff.). Global law overrides the core principles of domestic law: it is heterarchical and lacks the legitimacy of the law-making process established under democratic constitutions (Teubner 2000: 440). Teubner uses the example of *lex mercatoria* to explain the autonomy of global law, arguing that it develops strategies to resolve the paradox of self-validation (Teubner 1997: 10ff.).

This gives rise to the question of the collisions of norms and their resolution, arising from the overlap of legal regimes. Concepts such as legal unity, a hierarchy of norms, or the establishment of universal arbitration bodies cannot be employed for dealing with such collisions (Fischer-Lescano and Teubner 2004: 1003). According to Teubner and Fischer-Lescano, any attempt to establish a unity of global law is futile, since the fragmentation of global law reflects the fragmented global society, which is hardly a basis for the structuring of a normative legal system. Instead, legal orders must engage in decentralized networking in order to resolve the collisions of norms (Id.: 1017ff.).

Other authors, particularly Sousa Santos, emphasize the importance of the state as the central actor in global lawmaking, and further argue that legal pluralism is a hierarchical phenomenon (Sousa Santos 2002: 945).

Instead of legal systems existing autonomously, global legal pluralism is characterized by intertwined, interpenetrating, and interacting legal systems, indicating a "legal porosity or porous legality." State law may have lost the monopoly position it maintained under the prevalent notion of legal positivism, but it still plays the central role (Id.: 437–8).

When legal sociologists discuss the phenomenon of transnational law and the theoretical ramifications emanating from its existence, they usually consider it to be the product of civil society actors. *Lex mercatoria* is the prime example, upon which Teubner in particular expanded his theories. According to this perspective, transnational law – the legal order in-between international and national law – is a private governance phenomenon. This gives rise to the question of whether law at the transnational level can also be identified in areas where predominantly public actors, like government agencies or IOs, operate. Is there a stratum of rules, established by public actors, which is both distinguishable from and connected to other legal orders such as international and national law?

Four aspects suggest the existence of such a layer and its interconnection with other legal systems: the changing notion of sovereignty in the age of globalization, the growing role of soft law, the emergence of transnational bureaucracy networks, and the observations regarding transnational law. As pointed out above, the concept of sovereignty has evolved in response to the new challenges faced by the state. Today, it is no longer an impenetrable sphere. These changes become more apparent when one considers the emergence of supranational organizations and supranational law that directly affect member states. As sovereignty is less thought of as a principle to isolate the state and its legal order, the more likely are interconnections of the state's law to other legal systems.

Soft law is normally approached from a legal positivist perspective, and is thought to be created by IOs or states as a diluted or weaker version of supranational or international law. Although it has been identified as a possible governance instrument, its potential has not yet been fully explored. While prominent soft law instruments like UNGA resolutions or the informal agreements of large state conferences have been studied concerning their impact and interconnection with international and national law, other acts have been disregarded up to now. Presumably, there are other informal, less visible acts that flow from other public actors and interact with international or national law. In this regard, the role of transnational bureaucracy networks in the formulation and implementation of law has not yet been completely investigated. Their emergence indicates an opening up of traditionally closed structures with the goal of cooperative problem solving. When a state's administrative system opens up for informal modes of transnational cooperation, does this also make the national legal system open for informal legal acts, or – as put by Sousa Santos – does the porosity of the administrative system coincide with a porosity of the legal system?

The creation of law by civil society actors – transnational private law – demonstrates how such an informal legal system can emerge and interact with formal systems like international and national economic law. This suggests the existence of a type of informal law that shares the characteristics of transnational law like the *lex mercatoria*: an autonomous body of legal rules that interacts with other legal systems. The main difference is that the creators are public actors. Hence, in contrast to systems like *lex mercatoria*, it could be designated as transnational public law.

This type of law would be distinguishable from international law and supranational law. International law is the body of binding rules established by states to govern their relations. States create these norms and are themselves the main obligation holders. In some cases, the rules may also be directed at IOs. In order to become part of the national legal order, international law has to be incorporated by an act of law. Supranational law encompasses those rules issued mainly by supranational organizations like the EU, addressing not only states but also their citizens. It does not require an act of incorporation. Instead, it immediately becomes part of the national legal order. If transnational public law is to be thought of as a separate legal category it must also be distinguished from national law. All rules emanating from state institutions (i.e. the parliament or by parliamentary decision), or a government body like a ministry that governs the affairs of the population in a given area, make up the body of national law. The state is the sole creator and enforcer of the law, and the addressees are its denizens.

If law is approached functionally – as opposed to the formal approach taken by legal positivists – the preliminary typology of legal layers in the age of globalization is as follows in Table 5.1.

This classification reveals that the age of globalization is marked by the existence and interplay of several legal systems. Investigating transnational public law will complement the picture of law in this time, and its study can contribute to the solution of the theoretical conflict between the concepts of global legal pluralism and interlegality.

The typology in Table 5.1 is based on the assumption that creators of law are either public or private, and that while public actors can create

Table 5.1   Preliminary typology of legal layers in the age of globalization

Type	Creator	Addressees
Supranational law	IOs and states	States, societal actors
International law	States	States and IOs
Soft law	States and IOs	IOs, states, societal actors
Transnational private law	Societal actors	Societal actors
Transnational public law	Transnational bureaucracy networks	Public actors, societal actors
National law	State institutions	State institutions, societal actors

norms that address civil society actors, the converse situation, that is, that civil society actors create norms that apply to public actors, is not possible. Whether exceptions exist, for example, mixed forms of norm creation, has to be determined on the basis of empirical data.

Another important aspect of transnational public law is its legitimacy. This issue was briefly touched upon above in the discussion on the definition of law. It is often stated that authority is a feature inherent to law, and that such authority has to be legitimized. The matter of legitimacy also arises in connection with the disadvantages of transnational bureaucracy networks. As these play an important role in the creation of transnational public law, it is clear that the legitimacy of transnational public law could be problematic.

6
Conclusion

Changes in the international system initiated after World War II have intensified at a rate comparable to that of globalization, which has increased and grown more complex.

In terms of the role of the state in the international system, two things stand out. First, the concept of sovereignty has changed in such a way that it is no longer understood as impermeable or indivisible. Notwithstanding the continued recognition of sovereignty as an exclusive feature of the state, the conferral of sovereign rights to entities beyond the state is now acceptable. This essentially follows from the need for cooperative action to solve global problems. Second, areas that were, in the past, regulated by the state, now often fall outside its exclusive scope of governance. As long as national regulators operated within the administrative territory of the state, few difficulties arose in regards to governance. This is no longer the case, however, since national officials often have to work in a transnational setting, and the boundaries of national responsibility have been blurred. However, globalization forces the administration to loosen its territorial linkage and operate in settings beyond the state, jointly with administrative actors from other countries.

The importance of civil society actors, with their extensive command over governance resources, is slowly being recognized in international law. The growing role of NGOs and TNEs in governance arrangements has been observed – the aspects of governance both with the state and without have already been pointed out.

IOs have a long history, undergoing several changes in the past decades. However, some scholars have identified structural deficits that render them too inflexible to be an effective global governance tool. Instead, informal structures like transnational bureaucracy networks, the product of a transnationally operating administration, are put forward as substitutes. Transnational relations between administrative actors have been observed since the early 1970s, but have expanded rapidly in the past few decades.

The emergence and growth of transnational relations is characterized by the disaggregation of the state. However, the state does not disappear as the primary actor within the international system. Its subunits still exercise its power, albeit in a disjointed but flexible and adaptive fashion (Raustiala 2003: 10; 18–19). International law has also undergone profound changes. The notion of cooperation has become increasingly important, and is now the defining feature of international law. Treaty regimes have also been adapted to meet demands for flexibility. At the same time, other forms of law have gained significance. Soft law has already existed for quite some time, while recent observations indicate the emergence of a new type of law that is closely related to soft law, but also bears resemblances to the transnational law of civil society actors.

This brief summary points to a striking feature: transnational administrative relations, the extensive reliance on soft law, and the emergence of a public transnational law all indicate a high degree of informality in global governance structures. Such informality leads to a number of problems that will be discussed further in this book. The control of such networks and their activities is unclear, and since the exact form of transnational public governance remains unknown, a thorough empirical investigation of the actual measures applied in order to combat global environmental problems is necessary.

Part II

Solving Global Environmental Problems

7
Chemicals: A Global Challenge

The following sections describe the global problems caused by chemicals and examine the resulting legal challenges. The first section will provide an overview of the various global concerns arising from the ever-present use and release of chemicals. Following this will be a brief account of the toxicological methods used when addressing chemical safety. Finally, the difficulties finding legal solutions to the problems caused by chemicals will be discussed.

Chemicals as a global environmental problem

Chemicals[47] are ubiquitous. Virtually every industry is dependent on chemicals. The production process of almost all modern items – be it a pen, a computer screen or pair of jeans – is connected with chemical substances.[48] In the area of agriculture, innovative fertilizers and pesticides have made possible the immense increases in yields that were aptly dubbed the "green revolution" in the 1960s.

A few numbers reveal the economic importance of chemicals. The Chemical Abstracts Service (CAS) counts about more than 39,000,000 inorganic and organic substances (Chemical Abstract Service 2008a), and more than 25,000,000 commercially available substances (Chemical Abstract Service 2008b). The chemical industry manufactures goods worth US $1,600 billion annually (OECD 2008b: 3). In 2007, the trade

---

[47] When dealing with issues of chemical safety, several specific terms are used in toxicological or legal literature, for example, hazardous substance, hazardous material, dangerous substance, agent, etc. For the sake of simplicity here, the terms chemical, substance, etc. are used interchangeably.

[48] On the ubiquity of chemicals cf. CEFIC's campaign "Chemistry and You" (CEFIC 2008a).

in chemicals accounted for almost 11 per cent of world merchandise trade (WTO 2008: 43).

Yet the ubiquitous use has a downside: the chemicals industry can be considered a classic example of a technology that is both a solution to and an origin of problems. The complex production processes and the unwanted harmful effects of chemicals on the environment and human and animal health have lead to a "control problem" (Schneider 1989: 199–20). One aspect of the control problem stems from the complexity of the manufacturing process. This complexity holds the danger that either the harmful effects of the regular processes are neglected or ignored or failures in operational procedures can occur and lead to catastrophes. An example of the former is Minamata disease. Methylmercury was discharged into the Yatsushiro Sea, particularly the Minamata Bay area, by a plant producing various types of plastics. The substance accumulated in fish and shellfish, which were consumed by the local fishermen and their families. In 1956, first cases of a neurological syndrome appeared in Minamata. Only 12 years later was the disease attributed to severe mercury poisoning (National Institute for Minamata Disease 2008) and was linked to what appeared to be a "normal" production process.

There are also other, more drastic examples of the control problem such as manufacturing accidents, which have catastrophic effects on human health and the ecosystem. In this regard, Bhopal stands out as one of the worst industrial disasters ever: in 1984 methyl isocyanate leaked from a tank at a production plant, producing a cloud that killed several thousand people and injured many more. Another less severe accident occurred in Seveso in 1976 when dioxin was released after a chemical plant exploded. Although no one was killed or injured immediately, one long-term consequence was above-average rates of cancer and diabetes in this area (Nanda and Bailey 1989: 3–11; 17–19). The Schweizerhalle incident of 1986 occurred as result of a major fire. Large amounts of toxic chemicals drained into the Rhine, together with the water used to extinguish the fire, destroying the river's ecological system for hundreds of miles (Heil 1990: 11ff.).

The dangers posed by regular or accidental emissions being released during chemical production processes are less than those tied to the actual use of these substances (Scheringer 1999: 2). The control problem is increased as a result of the diffusion of risks and complexity of processes, in most cases with global and unforeseen consequences. Most prominently, the example of chlorofluorocarbons (CFCs) illustrates this issue. CFCs are almost unreactive and largely non-toxic, and their thermodynamic properties make them good coolants, widely used in coolers and air-conditioning systems. Since the mid-1970s, scientists had been warning of the danger of ozone depletion in the stratosphere; and in 1985, a hole in the ozone layer in the atmosphere over Antarctica was discovered (Wissenschaftlicher Beirat der Bundesregierung Globale Umweltveränderungen 2001: 28–31).

Another prominent example is the group of chemicals called chlorine compounds. Chlorinated hydrocarbons, such as Dichloro-Diphenyl-trichlorethane (DDT), pentachlorophenol (PCP), hexachlorobenzene (HCB), hexachlorocyclohexane (Lindane, γ-HCH), and polychlorinated biphenyl (PCB) are very versatile, and some are still being used for various purposes. DDT has been and is used as an insecticide; PCP as an algicide, fungicide, disinfectant, and preservative in the production of cellulose, paper, and cardboard; hexachlorocyclohexane for pest control; and PCP as a wood preservative. The applications for PCB are rather broad; it can be used as a lubricant or dielectric fluid for capacitors and transformers. HCB was used as a plant protection agent and softening agent in synthetics.[49] The abundance of these substances in the environment can be linked to their versatility and characteristic properties, including their environmental persistence, low water solubility, and ability to dissolve in fats. These mean their chemical structure does not break down under normal environmental conditions and they do not dilute and disperse in water, but instead accumulate in fatty tissue (Fiedler 2003: xi). As a consequence, these substances cause several environmental and health problems. Complex atmospheric processes – the interplay between evaporation and condensation – transport these substances on global air currents to the poles (the "grasshopper effect"). This means, for example, that DDT, used as an insecticide in equatorial regions, will ultimately accumulate in polar areas, far from its place of application. These substances accumulate in the fatty tissue of animals, with the highest concentrations being found at the top of the food chain, eventually affecting regional populations (Kallenborn and Herzke 2001: 216ff.). Consequently, a Canadian study found high levels of these persistent organic pollutants (POPs) in the breast milk of Inuit women in Canada's North.[50]

The actual toxicity of some of these substances has not yet been properly established. The results published in Rachel Carson's book, *Silent Spring*, in 1962, which caused a lot of public concern regarding the effects of the widespread use of pesticides, were gradually disproved as

---

[49] For details on the latter cf. Rippen 1987.
[50] Interview 17 October 2005: 3.

analytical capabilities improved. A large number of studies investigated the effects of exposure to DDT. Its metabolite dichlorodiphenyldichloroethylene methyl sulfonyl (MeSO$_2$-DDE) is regarded as a cause of breast cancer, but the environmental significance of these studies remains unclear (Zitkov 2003: 76–7).

Some of these substances are suspected of impairing the function of the endocrine system in animals, affecting, for example, oestrogen homeostasis – a process that self-regulates the production of reproductive hormones. So-called endocrine disruptors (EDs), which are exogenous substances that cause adverse health effects in intact organisms and their progeny, induce changes in endocrine function.[51] Some studies suggest that EDs, particularly DDT, play a role in the etiology of breast cancer by mimicking oestrogen (Brody et al. 2004; Safe 2004: 3ff.). Phthalates, one of the most widely used group of chemicals, also have endocrine disrupting potential. Despite their widespread use over the past 50 years, research regarding the harmful effects of phthalates has only recently intensified. Actual *in vivo* effects have not yet been fully explored (Harris and Sumpter 2001: 195ff.). Organotin compounds, such as tributyltin, which is used in antifouling paint for ship hulls, have deleterious effects on the endocrine systems of marine organisms. Tributyltin is linked with imposex, the imposition of male characteristics onto female sea snails, or intersex, the transformation of the female sea snail's palliale oviduct into a male prostate gland (Strand and Asmund 2003: 31ff.). Because of its bioaccumulative characteristics, high concentrations of tributyltin can be found in harbour mud, raising the problem of how to dispose of it, if a port basin has to be cleared (Brandsch et al. 2002: 139).

In addition to the transboundary conveyance by global airstreams, pesticides pose another global problem. Inconsistencies across regulatory regimes result in the prohibition of pesticides in one country, while they are still allowed in another. Moreover, it may be that the agent can be legally produced in the country where its application is prohibited and exported to the second. In this vicious circle, the pesticides may be "reimported" as residues in agricultural products.[52] These examples reveal a rather complex problem structure. From a toxicological perspective, it is very difficult to assess the actual interactions of substances in the environment; and, the cause and effect relations are sometimes

---

[51] Grünfeld and Bonefeld-Jorgensen (2004): 467–8; European Workshop on the Impact of Endocrine Disruptors on Human Health and Wildlife 1996.

[52] Cf. Ebbecke 2006: 18 for an overview of the problem.

diffuse. Transboundary effects, which span the globe and result from widespread use, contribute to the toxicological uncertainty.

Social and economic issues are a further component to this problem. Many developing countries are dependent on certain substances for pest control in agriculture. For example, DDT is an effective measure to combat Anopheles mosquitoes, the host animals and transmitters of Plasmodia sporozoites, which ultimately cause malaria and DDT is one of many substances that are severely restricted or banned in industrialized nations and at the same time exported to developing countries (Ross 1999: 499; Zahedi 1999: 708; 710–13). Often these countries lack the capacity to assess the hazards and risks a substance may pose properly; furthermore, the knowledge regarding the correct use of chemicals may be limited. Workers, in particular, are often unaware of potential hazards and basic safety precautions. As a result, they are often unnecessarily exposed to substances and their harmful properties. Many accidental exposures result from to improper storage, with the result that many farm workers in developing countries suffer pesticide poisoning (Ross 1999: 502ff.). Children, undernourished and often living in substandard hygienic conditions, are also affected by the misuse of hazardous substances. Ultimately, the exposure to toxic substances and poverty are directly related (World Bank 2002: 39), and the environment is also affected by their misuse. As a result of improper application or the use of outdated pesticides, the water and atmosphere are contaminated and wild animals endangered (Ross 1999: 504).

Maintaining chemical safety

Chemical safety consists of two phases. The first encompasses measures to identify the risks attached to the use of a specific substance. Hazards, exposure routes, and risks of a substance have to be determined, before measures aimed at averting harm to human health or the environment are employed. Therefore, a risk analysis needs to be carried out.[53] Measures aimed at mitigating or averting the risk form the second phase.

Identifying the hazards posed by a substance starts the first phase. The potential of a chemical to harm human health or the environment,

---

[53] The terminology is inconsistent. Sometimes risk analysis is used to describe the whole process – risk assessment, risk management, and risk communication. In other cases, risk assessment is synonymous for risk analysis, cf. Younes 2004: 46.

has to be investigated. The objective of hazard identification is to generate data regarding the intrinsic properties of a substance. This includes data on the substance's properties such as flammability, stability in water, and biodegradability. Quantitative tests on the dose-response relationship are carried out. This means, that tests are carried out to investivate, for example, the exact dose of a substance to be ingested orally to produce a toxic effect. All subsequent steps are based on the investigation of the substance's identified properties. Hazard investigation ultimately has the purpose of protecting human health and the environment (Klaschka, Lange and Madle 1997: 387). Consequently, the sensitivity of the test methods employed in the investigation of a substance's properties determines the efficacy of the risk management measures. A case in the field of pharmaceutical drug safety illustrates this. The drug Contergan contained the agent thalidomide and was placed on the market as a sedative and antiemetic for pregnant women without proper toxicological testing. Thalidomide, however, possessed teratogenic properties and eventually caused malformations of embryos (Spielmann 2004: 140; Hertel 2004: 432–3).

After the assessment, an exposure assessment is carried out. Possible emissions, dispersion pathways, and speed are investigated to assess the exposure of human health and the environment to the hazardous properties of a substance.

The subsequent risk assessment compares the chemical's hazardous properties, in particular, the findings of the dose-response analysis, to the number of likely exposures. In short, the "risk" pertains to the relationship between hazard and exposure. While a particular substance will always retain its hazardous properties, risk depends of a multitude of parameters. For example, a substance may be extremely volatile above a certain temperature. The risk of an accidental release is considerably lower in regions with a lower average temperature than in countries with a tropical or subtropical climate. Also important in this context are the possible uses of a substance. A chemical that is exclusively used as a catalyst in the production process – a so-called intermediate – can have extremely dangerous properties. However, if it is used in a controlled environment where only a few people are exposed to the substance, the risk that its hazardous properties will be a problem are lower than for a substance that is used as an active agent in pesticides used on crops.

Once the risks are known, scientists and politicians may begin to consider measures to minimize the risk. Risk management is not based solely on the scientific process of risk analysis, but is rather a political determination made in consideration of socioeconomic factors

such as the benefits arising from use of the substance and the societal acceptance of risks (Mahlmann 2000: 55; Ginzky 2000: 133ff.).
Possible measures are the prescription of maximum limits or restrictions or bans regarding the manufacture, placing on the market, or use
of a substance. Furthermore, substances can be allocated to predefined
categories based on the level of the hazard. The substance is classified
according to its hazardous properties and labelled with the corresponding signs and warning clauses to communicate its hazards and risks to
the users (Hertel 2004: 428–37).

National responses

Many national legal systems have responded to the problems caused
by the production and use of chemicals. First of all, production- and
product-related measures have been taken. The effects of the aforementioned catastrophes were far-reaching and drastic for the affected individuals. Nevertheless, the effects were locally or regionally confined.
Problems relating to the control of specific chemicals and their production can *ipso facto* also be addressed locally or regionally. For example,
the EC drew consequences from the disaster in Seveso and reacted with
regulatory measures to control the risks posed by accidents in certain
facilities.[54] Another production-related approach was the regulation of
the emissions resulting from chemical production, in cooperation with
which operators of installations like chemical or power plants must
apply the best available technique to reduce emissions of harmful substances (Pallemaerts 2003: 11).[55]

Finally, product-related measures implement the toxicological findings
of the risk assessment, thus they are ultimately based on toxicological
methodology. As early as the 1970s, product-related measures, including
legislative action, were already being taken in the EC and US, for example, with the prohibition of DDT.[56] Comprehensive regimes that had
the aim of putting toxicological methods of risk assessment into practice were introduced much later – in the late 1970s. In order to generate
the necessary data to carry out risk assessments, the US, EC, and Japan

---

[54] Council Directive 82/501/EEC ("Seveso I"), later replaced by Council
Directive 96/82/EC ("Seveso II").

[55] Cf. also Art. 9 paras 4, 10 and 11 Council Directive 96/61/EC.

[56] In Germany: *DDT-Gesetz*; in the USA: Cancelling of Federal registrations
of DDT products by the Administrator of the Environment Protection Agency,
cf. EPA 1972.

introduced extensive testing requirements.[57] One requirement was that new substances manufactured or placed on the market after the respective regime was introduced be tested for their physicochemical, toxic, and ecotoxic properties. As a result, information about hazards and risks is available for new substances. The investigation of the properties and risks of substance that have been on the market prior to the introduction of these testing regimes – so-called existing substances – falls under a different set of regulations. This legislation has only recently been introduced. For example, the EC introduced Regulation 793/93, a regulatory framework for the investigation of existing substances in 1993, 12 years after the investigation of new substances became mandatory. Further laws were created in the EC regulating the conduct of risk assessments,[58] and consequently, the imposition of restrictive measures, based on the previous scientific investigations and assessments.[59]

The existing laws have been proven to be rather ineffective – the investigation and assessment is a tedious, time-consuming procedure, and the data generated insufficient (Lahl and Tickner 2004: 161; Spieker gen. Döhmann 2003: 165–8).[60] In fact, there is a large gap in knowledge – aptly labelled "Toxic Ignorance."[61] In view of these toxicological uncertainties and the still-needed raw data, no risk assessments can be carried out, nor can restrictive measures be implemented.

In response to these legislative shortcomings, the EC has developed a new system that attempts to accelerate the risk assessment process without sacrificing a high level of health and environmental protection.[62] The system's acronym, "REACH," refers to its key features: registration, evaluation, and authorization of chemicals. Under the new

---

[57] EU: Council Directive 67/548/EEC, testing requirements for new substances were introduced by Council Directive 79/831/EEC, amending Directive 67/548; USA: TSCA; Japan: Law Concerning the Evaluation of Chemical Substances and Regulation of Their Manufacture etc. (Law No. 117); these laws are extensively analysed in Johnson, Fujie, and Aalders 2000: 341–71.

[58] Commission Directive 93/67/EEC.

[59] Art. 2a Council Directive 76/769/EEC.

[60] Cf. also Commission of the European Communities, Commission Working Document (SEC(1998) 1986 final), 8–14; United States General Accounting Office, Testimony Before the Subcommittee on Toxic Substances, Research and Development (GAO/T-RCED-94-263): 6.

[61] European Commission 1999; EDF 1997; EPA 1998a; cf. also European Commission, White Paper "Strategy for a future Chemicals Policy", COM (2001)88 final: 6.

[62] European Commission, White Paper "Strategy on a future Chemicals Policy," COM (2001) 88/final.

system, manufacturers and importers will have to produce data on all substances. There will be no distinction between new and existing chemicals. Furthermore, the chemical industry will be responsible for carrying out the costly risk assessments. The European Commission, assisted by the European Chemicals Agency, to be located in Helsinki, will be given the task of authorizing the use of certain chemicals, so-called substances of very high concern. REACH was adopted in late 2006 and entered into force 1 June 2007; specific sections of the regulation became applicable from 1 June 2008.[63]

Regulatory systems are intrinsically tied to the territory of the regulatory body. REACH, however, is exceptional in this regard. The adoption of REACH was preceded by extensive stakeholder dialogue. Of course, the European chemicals industry was heavily involved in this process, voicing its concerns. However, REACH also drew heavy criticism from the chemicals industry in the United States, protesting against the testing requirements for importers and downstream users.[64] This demonstrates the transboundary effects of a national or regional measure. It is conceivable that the impact of REACH on foreign chemicals industries will induce other countries to adapt their chemicals legislation accordingly (Winter 2007: 825–6).[65]

Its limited geographical scope notwithstanding, national legislation on the use of chemicals has aspects of global importance. One such element is the extent and depth of hazard investigation under national law. If the law does not require tests to investigate certain properties, for example, persistence or bioaccumulation, these remain unknown. The problem here is that the harmful impact of such properties does not necessarily happen in the area where the substance has been used. The same is the case when a chemical is tested for certain properties, but the required methods do not yield meaningful results for a thorough hazard or risk assessment. As a result, standardized testing procedures appear to be necessary. Collective standards have the additional benefit for the industry that test results are comparable. Findings can be used in different regulatory systems, thereby reducing the amount of required

---

[63] For details cf. Regulation 1907/06.

[64] Cf. ACC 2003; United States House of Representatives Committee on Government Reform – Minority Staff Special Investigations Division, A Special Interest Case Study: The Chemical Industry, the Bush Administration, and European Efforts to Regulate Chemicals.

[65] Cf. European Commission, Memo: Q and A on the new Chemicals policy, REACH (MEMO/06/488), item 18.

testing on vertebrate animals (Klaschka, Lange, and Madle 1997: 387; Koëter 2003: 13; Spielmann 2004: 141ff.).

A second element concerns the classification and labelling of substances. Global trade makes this aspect very important. The economic consideration here is that manufacturers and importers keep track of the various regimes so that they can classify and label their products accordingly. Differences in classification, which can result from different risk cultures or risk perceptions, lead to an uneven level of protection. If there is no such regime in place or if it is not properly enforced, which is often the case in developing countries, the labels from the state of origin remain, even if they are not readable or are incomprehensible for cultural reasons. Companies do little to address these issues. One interviewee described the situation in Thailand, where a German chemicals manufacturer produces pesticides for the domestic market and labels the products in Thai. When made aware of the fact that these products also appear on the Cambodian market, where no one understands the Thai warning labels, the company deemed this to be a problem of border control.[66]

"Toxic ignorance" amplifies the toxicological uncertainty, as no data is available to allow the proper conduct of risk analysis. The legal systems of individual states are overburdened with the task of generating the data on their own, which eventually affects the restrictions themselves. The political decision to restrict use of a chemical hinges on the outcome of the risk assessment. However, these restrictions cannot be carried out without proper data, and territorial limitations hinder their effectiveness. For example, there is no use in Sweden restricting POPs, if these substances are primarily produced in other states and used in Africa and, as a result of the "grasshopper effect," Sweden is affected by their dissemination, especially in the north of the country.

The global and multifaceted nature of the problem of the control of chemical dispersion suggests the need for international measures. Maintaining chemical safety – the safety of the products as opposed to the safety of their production – is essentially an international challenge and the solution of the control problem an international task (Hildebrandt and Schlottmann 1998: 1386; Alston 1978: 398–9; Wissenschaftlicher Beirat der Bundesregierung Globale Umweltveränderungen 2001: 28–31). Sharing the responsibility for the testing of chemicals, facilitating information exchange, and harmonizing chemical assessment procedures has economic benefits, as financial and human resources may be saved (Hiraishi 1989: 341).

---

[66] Cf. Interview (b) 14 October 2005: 30–1.

8
The System of International Chemical Safety

A multitude of actors is involved in international chemical safety, leading to numerous individual programmes and measures. For the sake of brevity and clarity, the following account will outline only the most important and relevant measures, which have the greatest impact on the problems described above.[67]

Actors

In view of the manifold uses of chemicals and the diverse dangers and diffuse risks posed by their application, it is clear that this issue falls within the mandate of several IOs. There is no single IO tasked with the widespread issue of chemical safety. Because of the varied uses and effects of chemicals, however, there are a large number of IOs whose fields of activity cover chemical safety issues such as workplace safety, health, the environment, or food safety participate in global activities. Activities carried out by of a large number of national authorities which participate in various related activities complement this work. Germany's chemical industry is one of the largest in the world and consequently, German authorities are quite experienced in chemical safety issues and are very active participants in international activities. Their involvement serves as an example for the large number of national authorities working in the area of international chemical safety. Several NGOs also participate in different programmes and activities, so that it is necessary to account for them, too, in this analysis.

---

[67] Cf. UNEP 2001 for a comprehensive overview.

IOs and their programmes

IOs play an important role in various activities. The following section will look at the most important IOs involved in issues of chemical safety in order to provide a picture of the institutional set-up regarding chemical safety. This will include a historical summary of the activities of IOs in the area of chemical safety.

OECD

When the OECD was set up in 1960 by the Convention of the OECD, as a successor to the OEEC, it was originally intended to operate as an economic counterpart to NATO.[68] Along with its geographic extension beyond the north Atlantic, the range of its activities has expanded beyond economic issues to include, for example, environmental matters. In recent years OECD enabled non-members and NGOs to participate in its activities, and, as a result, OECD became a meeting place primarily for civil servants. In comparison with other IOs, it does not produce a significant amount of hard law, but heavily relies on soft law and compliance through dialogue, peer pressure, and threats of loss of reputation (Marcussen 2004: 103; 112).

The development and implementation of chemical programmes involves a number of institutions within the OECD structure. The OECD's supreme decision-making body is the Council (Art. 7 of the OECD Convention). Representing the will of OECD member states, all decisions made within the organization emanate from the Council.

The supreme body at the working level is the Secretariat, led by the Secretary-General (Arts 10 and 11 of the OECD Convention), responsible to the Council. An Environment Directorate, undertaking the Secretariat's environmental programme of activities, was established in 1970.

In 1970, the Council established the Environment Policy Committee (EPOC), whose mandate was recently renewed.[69] EPOC's responsibility is to implement the environmental aspects of the OECD's work programme. It oversees four working parties, among them the Working Party on Chemicals, Pesticides, and Biotechnology (OECD 2005: 5). This working party was first established in 1971, and EPOC renewed its mandate in 2004. Most of its tasks are carried out jointly with the

---

[68] On the origins of OECD, cf. Hahn 1997: 791–2.
[69] OECD Council Resolution C (2004) 99/REV1.

Chemicals Committee, which was set up in 1978 by the Council.[70] In some cases, the same person represents a member state in EPOC and the Chemicals Committee as responsibilities of the two bodies overlap. The key difference between EPOC's Working Party and the Chemicals Committee is funding: the working party is funded by all OECD member states, whereas the Chemicals Committee receives its funds from a budget provided by those member states actively involved in this specific programme.[71] For example, in 2006 Germany advanced € 225,000 to the OECD chemicals programme.[72] These funds were provided in addition to the € 28.8 million contributed by Germany to the overall OECD budget.[73]

EPOC and the Chemicals Committee together form the Joint Meeting, which carries out and supervises the implementation of specific programmes, as well as identifying problems and elaborating policies. In order to perform its various tasks, the Joint Meeting has established and oversees a number of working groups, the most relevant here being the Working Group of National Co-ordinators of Test Guidelines Programme (WNT).[74] Several other subsidiary bodies also exist, such as the Task Force on Existing Chemicals (Task Force), which is responsible for the OECD programme on high production volume (HPV) chemicals.

The meetings of these bodies are open to observers: the Business and Industry Advisory Committee (BIAC), the Trade Union Advisory Committee (TUAC), IOs, NGOs, and non-member states. For example, several IOs, such as the Council of Europe, UNECE, WHO, WTO, and others are present as observers in EPOC; Israel and Slovenia are observers in the Chemicals Committee.

OECD's Environment, Health and Safety Programme (EHS) provides the programmatic framework for these units. The EHS Programme is part of OECD Environment Programme adopted by the Council. It is the framework for all activities related to chemical safety, emerging from a

---

[70] OECD Council Decision C (78) 127 (Final), the mandate has been renewed regularly.

[71] Cf. Art. 4 of Part II OECD Council Decision C (78) 127 (Final).

[72] Cf. Bundeshaushaltsplan 2006, Einzelplan 16, Kapitel 2, Titel 687 03-332, item 2.

[73] Cf. Bundeshaushaltsplan 2006, Einzelplan 60, Titel 687 22-022.

[74] Cf. 39th Joint Meeting of the Chemicals Committee and the Working Party on Chemicals [ENV/JM/M(2006)1, Annex II], where the tasks and objectives of the WNT are set out.

chemicals programme set up in 1971 (OECD Secretariat 2008b: 7).[75] In general, the EHS Programme promotes the harmonization of national approaches to chemical safety, developing common tools that are later implemented by OECD countries and some non-member states. Furthermore, it provides states with a forum to exchange views and information (de Marcellus 2003: 125–6).

The EHS Programme consists of 12 subprogrammes, two of the most relevant ones, the Test Guidelines Programme and the Existing Chemicals Programme, will be discussed in more detail below. In general, the OECD's main objective is the harmonization of legal requirements in its member states to avoid the distortion of competition. Thus, the initial rationale for the OECD's chemicals programmes was economic, rather than environmental (Schneider 1988: 98; 193). However, it is clear that economic and environmental aspects are inextricably linked with each other, as the chemicals example vividly demonstrates. As 80 per cent of the chemicals produced worldwide are manufactured in OECD member countries, the OECD is an important actor in the development of international chemical safety (Gärtner, Küllmer, and Schlottmann 2003: 4605).

The consensus principle set out in Art. 5 of the OECD Convention requiring binding decisions to be adopted unanimously, leads to a weakness of OECD in stipulating binding instruments of chemical safety, as an "industry-friendly" country might veto the adoption of an allegedly disagreeable act. As one expert proclaimed, the OECD is "too soft."[76] The OECD is therefore not the place to develop incisive legally binding instruments. However, the OECD does provide a forum for the representatives of countries responsible for the bulk of the world's chemical production. The fact that the OECD is made up of only 30 countries (instead of more than 100 like other IOs) could be considered an advantage, in terms of   both its flexibility and the speeding up of the decision-making process.

For the promotion of international chemical safety, the OECD relies on "soft pressure," such as, for example, the Mutual Acceptance of Data (MAD) system of Test Guidelines and Good Laboratory Practice (GLP) principles. Implementing this system is not obligatory, but it makes sense from an economic perspective.[77]

---

[75] The Council addressed the issue of chemical safety for the first time in OECD Recommendation C (71) 83/Final.

[76] Interview 11 July 2002: 5.

[77] Cf. 153 below for details on MAD

UN Specialized Agencies and other UN institutions

Several Specialized Agencies within the UN framework carry out individual or joint programmes in the field of chemical safety. In addition, several other institutions such as the United Nations Environment Programme (UNEP) or the United Nations Institute for Training and Research (UNITAR) operate in this area.

The World Health Organization (WHO) was established in 1948 with the entering into force of the WHO Constitution, its objective being according to Art. 1 of the WHO Constitution the "attainment by all peoples of the highest possible level of health." The WHO's supreme decision-making body is the World Health Assembly (WHA), made up of representatives from the member states. Among other functions, it has the task of determining the policies of the WHO and establishing subsidiary bodies to carry out specific activities (Art. 18 of the WHO Constitution). The Executive Board serves as the executive body of the WHA (Art. 28 of the WHO Constitution). Administrative tasks are administered by the Secretariat, and headed by the Director-General (Arts 30–7 of the WHO Constitution). The WHO has been concerned with the effects of hazardous substances, particularly pesticides, since the 1950s. It has a long history of cooperation with the Food and Agriculture Organization (FAO) and the International Labour Organization (ILO) (Schneider 1988: 97; 189; Mercier 1981: 39–40). Noteworthy is, for example, the Joint FAO/WHO Meeting on Pesticide Residues, which has been in operation since 1961 (Mercier 1981: 39). The WHO's Guidelines for Drinking Water also touch upon issues of chemical safety.[78]

The ILO was originally founded in 1919; the Treaty of Versailles provided in Art. 387 for the establishment of an organization to promote humane labour conditions. The ILO is founded on the Constitution and the Declaration of Philadelphia from 1944. The International Labour Conference is ILO's decision-making body. It is made up of the member state representatives who determine its policies. According to Art. 19 ILO of the Convention, the Conference can adopt conventions and recommendations. Member states are obliged to present conventions to the competent national authorities for ratification within a year of their adoption. Since its establishment, the ILO has been active in the field of international chemical safety, issuing recommendations concerning

---

[78] Guidelines for Drinking Water Quality, Chapter 8 – Chemical Aspects, Third ed, 2003.

the handling of white phosphorous or lead in its founding year.[79] A Convention on the use of lead in paint followed in 1921,[80] with several general conventions and recommendations regarding workplace-related issues of chemical safety following.

UNEP is not an IO in the strict sense, but rather a UNGA programme carried out within the framework of the UN. The origins of UNEP can be traced back to the United Nations Conference on the Human Environment (UNCHE), which took place in Stockholm in June 1972 (Kilian 1987: marginal ns 1–2; Kilian 1987: 235–53). It was established through a resolution of the UNGA in 1972 (UNGA Res. 2997 (XXVII)), which defined the budget, objectives, and set-up of UNEP. UNEP's objective was the implementation of recommendations set out in the Action Plan adopted at the UNCHE (Gray 1990: 294). Its scope is comprehensive, taking a cross-sectoral approach rather than dealing with environmental problems from a specific (for example, health- or work-related) perspective. The organizational structure of UNEP is modelled on the structure of IOs such as the WHO and ILO. UNEP's principal organ is the Governing Council, whose members are elected by UNGA. The Governing Council directs UNEP's general policy and supervises the Secretariat and the Environment Fund. UNEP's executive organ is the Secretariat, which carries out the decisions of the Governing Council. Finally, the Environment Fund bears the costs of the implementation of the various environmental activities within the UN system (Id.: 295–6; Kilian 1995: marginal n. 12–17).

UNEP has dealt with the issue of chemicals since the mid-1970s. Following Recommendation 74 (e) of the Action Plan for the Human Environment adopted by the UNCHE, the International Register of Potentially Toxic Chemicals (IRPTC) was established at UNEP in 1976. The IRPTC collects and disseminates data on the production volumes and properties of chemicals, mainly by compiling and linking information already accumulated in various national or regional systems (Wagner 1998: 245; Huismans 1980: 393–403), thereby facilitating access to existing scientific data (Alston 1978: 419). It also maintains a legal file for information on international and national policies and regulations regarding the handling, transport, storage, disposal, and use of substances (Keita-Ouane et al. 2001: 112). It does not, however, proactively warn countries of the deleterious properties of particular

---

[79] White Phosphorus Recommendation (No. 6).

[80] White Lead (Painting) Convention (C13) and Lead Poisoning (Women and Children) Recommendation (No. 4).

substances (Pallemaerts 2003: 442). In order to remedy this problem, UNEP took up another function in 1994: the improvement of information exchange procedures. In 1987, the Governing Council adopted the London Guidelines for the Exchange of Information on chemicals in International Trade, introducing a voluntary prior informed consent (PIC) procedure for certain substances (Id.: 445ff.).[81]

In 1995 UNEP changed the name of its chemicals programme from IRPTC to UNEP Chemicals, indicating that it should take a more comprehensive approach to chemical safety instead of limiting it to the collection of data (Wagner 1998: 247). This was underlined by the Governing Council's decision charging UNEP with the development of an international, legally binding notification procedure for certain dangerous chemicals and pesticides in international trade as well as an international convention for the protection of human health and the environment from POPs (Merkel 1997: 133–4).[82] Today, UNEP Chemicals is UNEP's center for all activities regarding international chemical safety (UNEP Chemicals 2001: 117). Although the range of UNEP's activities relating to chemical safety is broader,[83] this investigation will concentrate on UNEP Chemicals.

The FAO was founded in 1945.[84] It is a specialized UN organization with the purpose of raising living standards and increasing the availability of agricultural products (Preamble and Art. 1 of the FAO Constitution). It is concerned with chemical safety in two ways: first, it maintains programmes on the proper application of pesticides to increase yields and control pests (FAO 2008) and second, it is concerned with food safety, establishing, together with the WHO, the Joint Meeting of the FAO Panel of Experts on Pesticide Residues in Food and the Environment and the WHO Expert Group on Pesticide Residues, bringing together government experts in order to formulate maximum residue limits. Finally, these limits are submitted to the Codex Alimentarius Commission.

The United Nations Industrial Development Organization (UNIDO) was created in 1966 with the aim of promoting and accelerating industrial development (Art. 1 of the Constitution of UNIDO). By 1967, UNIDO had set up a Chemicals Industries Branch to provide

---

[81] UNEP Governing Council Decision 14/27.

[82] UNEP Governing Council Decisions 19/13 A–D.

[83] Including the protection of the Ozone Layer, chemical accidents, marine protection, and biodiversity; cf. UNEP 2001: 117–19.

[84] Cf. Schütz 1995 for a detailed review of the FAO.

support to developing countries seeking to build indigenous chemicals industry capacity. The accident in Bhopal caused a shift towards chemical safety. Since 1992, UNIDO has been involved in the implementation of the Montreal Protocol on Substances that Deplete the Ozone Layer, helping developing countries to phase out their industrial use of ozone-depleting substances (UNEP 2001: 151). As the result of structural reform in 1998, most activities pertaining to chemical safety operate within the Cleaner Production and Environmental Management Branch. A separate branch is dedicated to the Montreal Protocol (UNIDO 2008b).

UNITAR was established in 1965 by the United Nations Secretary General in accordance with an UNGA resolution.[85] UNITAR is an autonomous institution within the UN framework (Art. 1 of the Statute of UNITAR). According to Art. 2 of its Statute, UNITAR has, *inter alia*, the function of providing training at various levels to persons (particularly from developing countries) for assignments connected with the UN or its specialized agencies. In the field of chemical safety, the institute maintains its Training and Capacity Building Programme in Chemicals and Waste Management (CWM). The objective of CWM is to support developing countries and countries in economic transition in the management of chemicals (UNITAR 2008). UNITAR carries out a number of programmes aimed at facilitating national infrastructure assessment, strategy development, and integrated chemicals management, and specialized training and capacity building programmes (UNEP 2001: 155–9).

The UN and its Economic and Social Council

The Economic and Social Council (ECOSOC) of the UN is another important player in the field of international chemical safety.[86] Art. 1 (3) of the UN Charter defines as one of the objectives of the UN "achiev[ing] international co-operation in solving international problems of an economic, social, cultural, or humanitarian character, and in promoting and encouraging respect for human rights and for fundamental freedoms for all without distinction as to race, sex, language, or religion." This is rendered more precisely in Art. 55(b) of the UN Charter: the United Nations shall promote "solutions of international economic, social, health, and related problems; and

---

[85] UNGA Resolution 1934 (XVIII); cf. Rittberger 1995 for an introduction to UNITAR.

[86] A detailed account of ECOSOC is provided by Lagoni 1995.

international cultural and educational cooperation...." The task of taking measures to tackle these issues devolves to ECOSOC (Art. 60 of UN Charter).

ECOSOC exercises most of its functions through subsidiary bodies (Lagoni 1995: marginal n. 6), which were established as such in 1953 with the UN Committee of Experts on the Transport of Dangerous Substances (UNCETDG)[87]. The task of one subsidiary body is the institution and further development of the Recommendations on the Transport of Dangerous Goods (UNRTDG). The UNRTDG were first published in 1957 and are constantly revised; since 1996 Model Regulations have been included to facilitate the adoption of the UNRTDG[88]. Although not legally binding, the UNRTDG serves as the basis for a large number of international treaties and national laws regulating the transport of hazardous materials.[89]

The Committee is affiliated to the United Nations Economic Commission for Europe (UNECE), one of the ECOSOC's regional commissions set up pursuant to Art. 68 of the UN Charter.[90] UNECE services the UNCETDG, providing secretarial functions.

The organizational structure was rearranged in 1999. As the Globally Harmonized System of Classification and Labelling of Chemicals (GHS) was conceived as a dynamic system, a special committee was established. Today, the Committee of Experts on the Transport of Dangerous Goods and on the Globally Harmonized System of Classification and Labelling of Chemicals (UNCETDG/GHS) operates at a strategic level and supervises the Subcommittee of Experts on the Transport of Dangerous Goods (UNSCETDG) and the Subcommittee of Experts on the Globally Harmonized System of Classification and Labelling of Chemicals (UNSCEGHS); both conduct the real work at a technical level.[91] The supervisory work comprises the functions of approving work programmes or formal endorsement and submission of work results. But UNCETDG/GHS is not supposed to intervene in the

---

[87] ECOSOC Resolution 468 G (XV).

[88] The current edition is the 13th: United Nations Recommendations for the Transport of Dangerous Goods/Model Regulations, Geneva, 26 April 1957, 13th revised edition, New York and Geneva 2003 (ST/SG/AC.10/1/Rev.13). The 14th edition has been presented to ECOSOC for adoption, Draft Resolution 2005/... (UN/CE TDG-GHE/2/INF.2.

[89] *Infra* 100.

[90] Cf. ECOSOC Resolution 36 (IV).

[91] ECOSOC Resolution 1999/65; United Nations Secretariat, Note by the Secretariat (ST/SG/AC.10/C.4/2001/7), Annex II.

decision-making process of the subcommittees, nor to review or even alter recommendations agreed upon by subcommittees.[92] The member states who sit on each committee are elected by ECOSOC.[93] While UNSCETDG retains the task of developing the UNRTDG and Model Regulations, UNSCEGHS is responsible for the implementation and revision of GHS.[94] Experts from various UN member states convene in the UNSCEGHS and only they have the right to vote.[95] UN member states can apply for membership, but the composition is decided by ECOSOC.[96] IOs participate in the activities, but may not vote. NGOs maintain a consultative status.[97]

Interorganizational and intergovernmental institutions

Three entities exist within the international chemical industry which prima facie are not IOs but due to their organizational set-up and relevance cannot simply be categorized as subsidiary bodies or working programmes of the aforementioned institutions. For the time being they will be defined as Interorganizational and Intergovernmental Institutions.[98]

The International programme on chemical safety

Considering the many IOs maintaining Programmes that address various aspects of international Chemical Safety, it is not surprising and rather probable that activities will overlap. This results in an unwanted duplication of work and is as such a waste of resources.

This led the WHA to request the WHO Director-General to examine possible options for international cooperation to address issues surrounding the toxic effects of chemicals.[99] His report was later endorsed

---

[92] Cf. Para 1 Draft Terms of Reference laid down in the Annex to ECOSOC Resolution 1999/65.

[93] UN Secretariat, Note by the Secretariat (ST/SG/AC.10/C.4/2001/8); ECOSOC, Organizational Session for 2001, (E/2001/L.2/Add.1). Consequently, its members do not come from only European countries.

[94] Chapter 1, Section 1.1.3.2.1 GHS.

[95] United Nations Secretariat, Note by the Secretariat (ST/SG/AC.10/C.4/2001/1) in conjunction with Rules 58 and 27 Rules of Procedure of the Economic and Social Council.

[96] Para. 3 ECOSOC Resolution 1999/65.

[97] United Nations Secretariat, Note by the Secretariat (ST/SG/AC.10/C.4/2001/1) in conjunction with Rule 79 and Rules 80ff. Rules of Procedure of the Economic and Social Council.

[98] Chapter 11 will analyze the status of these institutions in more detail.

[99] WHA, Evaluation of the Effects of Chemicals on Health (WHA 30.47).

by the WHA, which called upon him to promote international cooperation further and strengthen the implementation of the chemicals
programme through a central WHO unit and a network of national
institutions.[100] Finally, the Executive Board endorsed the plans for an
international programme in the shape of a collaborative effort of WHO,
ILO, UNEP, and national authorities.[101] The International Programme
on Chemical Safety (IPCS) was thus not conceived as an entirely new
approach but rather was designed to strengthen existing activities and
jointly initiate new ones (Mercier 1981: 41).

Hence in 1980, WHO, ILO, and UNEP concluded an MoU to establish the IPCS with the goal of bundling or better coordinating their
own activities in the field of chemical safety (Id.: 39; Schneider 1988:
98).[102] The MoU has been repeatedly extended and in 1996 partly
revised to incorporate developments since the adoption of Agenda
21,[103] the most important of these being the designation of the IPCS
as the centre of international cooperation in the field of chemical
safety.[104]

Since the inception of the IPCS in 1980, 36 countries have agreed to
participate in its activities (Hildebrandt and Schlottmann 1998: 1389).
Relations between the countries and IPCS are regulated by MoUs.[105]
Participation also entails financial contributions. For example, in 2006
Germany funded the IPCS with € 584,000.[106]

According to the original MoU, the IPCS is furnished with an appropriate organizational structure to fulfil its objectives.[107] It is made up of
the Intersecretariat Coordinating Committee (ICC), the Central Unit

---

[100] WHA, Evaluation of the Effects of Chemicals on Health (WHA 31.28).

[101] Executive Board, Evaluation of the Effects of Chemicals on Health
(EB.63.R.19).

[102] Memorandum of Understanding between the United Nations Environment
Programme, the International Labour Organization and the World Health
Organization, Concerning Cooperation in the International Programme on
Chemical Safety (MoU IPCS).

[103] ILO Governing Body, Entry Into Force of the Revised Memorandum of
Understanding Concerning Cooperation in the International Programme on
Chemical Safety (UNEP, ILO, WHO) (GB.268/LILS/4/1), para. 1.

[104] Chapter 19, para. 6 of Agenda 21.

[105] For example, the MoU regarding U.S. Environmental Protection Agency
collaboration in the IPCS; Cf. United States Department of State 2007: 312.

[106] Cf. the Bundeshaushaltplan 2006, Einzelplan 16, Kapitel 02, Title 687
03-332 item 4.

[107] Para. 13 of the IPCS MoU.

(CU), the Programme Advisory Committee (PAC), Task Groups, and Working Groups.

The decision-making body of the IPCS is the ICC, composed of representatives from the Cooperating Organizations, that is, UNEP, ILO, and the WHO. The ICC decides on the activities the IPCS will undertake, approves work plans, and provides guidance to the IPCS Director regarding the implementation of recommendations made by the PAC.

The CU is responsible for the management and coherence of the IPCS. It operates under the control of the Director, who is appointed by the Director-General of the WHO in consultation with the executive heads of the Cooperating Organizations. Besides administrative duties, the CU undertakes scientific and technical work on behalf of the Cooperating Organizations.

The PAC provides advice concerning various aspects of the work undertaken within the framework of IPCS. It serves as an advisory body and consists of 20 experts, 18 of whom are appointed by the Director-General of the WHO in consultation with the Participating Organizations; the remaining two are an employer and a worker appointed by the ILO Governing Body.

In addition to these bodies, the IPCS is supported by a large network of so-called Participating Institutions (PI) comprised of national, regional, and international institutions with governmental, intergovernmental, or non-governmental mandates.

National Focal Points in the participating countries are supposed to disseminate information from the IPCS in their respective country and also relay the country's views back to the IPCS (IPCS 2003; Mercier 1981: 41–2).[108]

The WHO supplies the bulk of IPCS funding and other resources and thus has a prominent role. The ILO contributes a comparatively small portion of the budget and UNEP does not contribute at all. UNEP nevertheless plays an important role by offering necessary environmental perspectives to IPCS activities, whereas the ILO and WHO both focus on human health issues.[109]

As a result of its exact structure according to the tripartite arrangement, the IPCS enjoys more leeway than similar WHO divisions. One expert explained that "[the IPCS] can tell WHO, that…that this [a particular activity] is something that is needed from the point of view of all these three organizations. And [the IPCS] need not necessarily only

---

[108] Paras 11–13, Annexes I, III, and IV of the IPCS MoU.
[109] Interview (a) 14 October 2005: 6–8.

listen to…to…to one organization only."[110] Advice from outside experts does play an important role in the decision-making process, but the IPCS is otherwise relatively autonomous in its choice of projects, with its ultimate limits determined by funding.

The IPCS has two objectives, first, to establish as its normative function scientific foundations for assessing risks and safe use of chemicals and second, to further technical cooperation.[111]

To fulfil its objectives, the IPCS maintains many programs, including the publication of reports on substances (Environmental Health Criteria) or their respective risk assessments (Concise International Chemical Assessments, CICADs) (IPCS 2003: 12, para. 54; Gärtner, Küllmer, and Schlottmann 2003: 4604). Beneficiaries of work done within the framework of the IPCS are not just the collaborating states, but also developing countries, which often do not maintain the capacity to assess chemicals and their risks properly (Id.: 4604).

The IPCS maintains a website (www.inchem.org), through which it can quickly disseminate the information compiled in its various reports and documents, making access to technical documents such as substance reports easy and convenient. Political documents such as the results of PAC meetings or the content of MoUs are not available online.

Inter-Organization Programme for the Sound Management of Chemicals

To coordinate their activities for implementing the UN's Agenda 21, Chapter 19 UNEP, WHO, ILO, FAO, UNIDO, and OECD established the Inter-Organization Programme for the Sound Management of Chemicals (IOMC) in 1995 (Hildebrandt and Schlottmann 1998: 1387). UNITAR joined the IOMC in 1998 (IOMC 2008a). The World Bank and the United Nations Development Programme also participate as observer organizations.[112] In 2001, the International Maritime Organization (IMO) was invited to join the IOMC.[113] The IOMC also informally collaborates with the Organisation for the Prohibition of Chemical Weapons on a case-by-case basis.[114]

---

[110] Interview (a) 14 October 2005: 10–11.

[111] Para. 8 of the IPCS MoU.

[112] IOMC IOCC, Summary Record of the Eighteenth Meeting (IOMC/IOCC/03.44).

[113] IOMC IOCC, Summary Record of the Fourteenth Meeting (IOMC/IOCC/01.01 Rev.1), Item 5.

[114] IOMC IOCC, Summary Record of the Thirteenth Meeting, (IOMC/IOCC/00.08), Item 5.

The IOMC is based on an MoU between the participating IOs.[115] According to the MoU, the purpose of the IOMC is to coordinate the common or individual policies and activities of the participating organizations. The activities match the six programme areas of Agenda 21, Chapter 19, but are not limited to them.[116]

The IOMC has two organs: the Inter-Organization Coordinating Committee (IOCC) and a Secretariat. The IOCC is composed of one representative from each of the seven participating organizations. It adopts its own rules of procedure and elects a Chairperson and, if necessary, a Vice-Chairperson. The IOCC meets at least twice a year and may invite observers and set up advisory bodies. The main functions of the IOCC are to coordinate and align the activities of the Participating Organizations to prevent work overlap and promote joint programmes.[117] A Secretariat is set up to provide the IOCC with organizational services and is located at the WHO, its administering organization.[118]

Technical Coordinating Groups (TCGs) have been set up at the technical level, which aim to enable consultation between the participating organizations (IOMC 2008b). The IOCC has issued Standard Operating Procedures (SOP) for the TCGs. According to the SOP, depending on the agreement of the IOCC, the groups may invite representatives from intergovernmental organizations, governments, and international industry, labour, and public interest NGOs, provided they are active in the relevant area. The TCG on the Assessment of Existing Industrial Chemicals and Pollutants serves a special purpose as it helps to coordinate IPCS and OECD programmes in this area and thus contributes to the prevention of duplication of work, even though this may be impossible to eliminate completely (IPCS 2003: 12, para. 50–1).

Intergovernmental Forum on Chemical Safety

When it became clear during the preparation of UNCED that the IPCS should be designated as the centre for international cooperation on chemical safety, the UNCED Preparatory Committee invited UNEP, the ILO, and the WHO to report on ongoing work. In 1991, a meeting was held with government experts in London to discuss the enhancement of international cooperation on chemical safety, recommending among

---

[115] MoU Concerning the Establishment of the Inter-Organization Programme for the Sound Management of Chemicals (IOMC MoU).
[116] Para. 2 of the IOMC MoU.
[117] Paras 3ff. of the IOMC MoU.
[118] Para. 7 of the IOMC MoU.

other things the establishment of an intergovernmental forum on chemical risk assessment and management. Reference to this meeting was made in Chapter 19, para. 76 of Agenda 21, which invited UNEP, the ILO, and the WHO to convene a meeting on chemical safety to consider further the 1991 recommendations. In April 1994 in Stockholm these organizations convened the International Conference on Chemical Safety (ICCS) (Carpenter and Krueger 1997: 1–2). Representatives from 114 countries attended the ICCS, as well as UN bodies, UN Specialized Agencies, and several other intergovernmental organizations and NGOs.[119] One of the most prominent results of the ICCS was a resolution on the establishment of an Intergovernmental Forum on Chemical Safety (IFCS).[120]

Conceived as a worldwide forum, taking place biennially or triennially,[121] the IFCS has developed strategies for the implementation of Chapter 19 of Agenda 21, encouraging collaborative efforts among key actors in this area.[122] Participation is open to governmental participants, intergovernmental participants, and non-governmental participants. While non-state actors may attend as observers at the Conference of the Parties (CoP) to some conventions or at the sessions of an IO's decision-making body, the IFCS ToR explicitly provide for these groups to bring forward their ideas and suggestions and participate more actively. Thus, the IFCS enables the relevant actors in the field of international chemical safety to meet almost on equal terms.[123] But, only the governmental participants have the right to vote. All decisions of the IFCS are to be reached by consensus. However, if this is not possible, decisions may be taken by a simple majority.[124] In practice, however, majority decisions do not occur and all decisions of IFCS have been reached through consensus.[125]

Forum sessions are managed by the President and five Vice-Presidents elected at each session from among the governmental representatives,

---

[119] The International Conference on Chemical Safety, Final Report, Para. 1.

[120] Resolution on the Establishment of an Intergovernmental Forum on Chemical Safety (IPCS/ IFCS/ 94.Res.1); changed and amended at IFCS III, Final Report (IFCS/ Forum III/ 23w).

[121] Para. 5 of the IFCS ToR; 1994: IFCS I, Stockholm; 1997: IFCS II, Ottawa; 2000: IFCS III, Salvador da Bahía; 2003: IFCS IV, Bangkok; 2006: IFCS V, Budapest; 2008: IFCS VI, Dakar.

[122] Paras 1–2 of the IFCS ToR and IFCS III, Final Report (IFCS/ Forum III/ 23w), Annex I.

[123] Interview b) 14 October 2005: 6–7.

[124] Para. 12.1 of the IFCS ToR.

[125] Interview b), 14 October 2005: 9–10; Paras. 1.2 and 14. 2 of the IFCS ToR.

as well as the Forum Standing Committee (FSC).[126] The President's task is to represent impartially all participants at the forum and act as its Chairperson, ensuring observance with the IFCS ToR.[127] Promoting the IFCS and its recommendations and fostering cooperation in their own regions are functions of the Vice-Presidents.[128]

The composition of the FSC is very heterogeneous, consisting of the IFCS elected officers and IOCC, government, and NGO representatives.[129] The FSC assists in the preparation of the coming forum and, as its members voice the views of their constituency, provides early input on new issues to be considered by the forum (UNEP 2001). The forum or the FSC may establish ad hoc working groups for specific tasks.[130] These are made up by governmental representatives but may be open to other participants who, however, do not have the right to vote. Each state maintains National Focal Points to coordinate and disseminate information on IFCS activities in their own countries. They report to the IFCS Secretariat annually on their implementation of IFCS Priorities for Action and other matters of chemical safety.[131] Secretarial services for the IFCS are provided by the WHO.

Off the record, an international lawyer once described the IFCS as a "strange being." Indeed, it is noteworthy, that the term "conference" was abandoned in the ToR adopted at the ICCS and replaced with "forum," stressing its informal character. In fact, the IFCS is devised as a "non-institutional arrangement,"[132] declaring that it is not another IO. The reason behind this choice was to prevent the IFCS from operating as an organization (in, for example, developing its own activities), and thus competing with other IOs for the scarce funds available for activities in the field of international chemical safety.[133] In 2008, the budget of the IFCS' Secretariat amounted to some € 510,000, of which Germany provided 20 per cent.[134]

Since the adoption of the Strategic Approach to International Chemicals Management (SAICM) in 2006, IFCS faces an uncertain future. At IFCS

---

126 Annexes II and III of the IFCS ToR.
127 Annex II of the IFCS ToR.
128 Annex III of the IFCS ToR.
129 Para. 7.2 of the IFCS ToR.
130 Para. 8 of the IFCS ToR.
131 IFCS III, Final Report (IFCS/Forum III/23w), Annex 4.
132 IFCS III, Final Report (IFCS/ Forum III/ 23w), Art. 1 para. 1 Annex I.
133 Interview b) 14 October 2005: 11–12.
134 Bundeshaushaltsplan 2008, Einzelplan 16, Kapitel 02, Titel 687 01-332, item 13.

VI, which took place in Dakar in 2008, participants undertook the first step towards defining the relationship between the IFCS and SAICM. The "Dakar Resolution on the Future of the Intergovernmental Forum on Chemical Safety (IFCS)"[135] stresses in its preamble the "unique multifaceted role that the Forum has played as a flexible, open, and transparent brainstorming and bridgebuilding forum..." and suggests that the International Conference on Chemicals Management (ICCM) integrates the IFCS into its structure, whereby the exact role of IFCS within SAICM would yet have to be determined. This process of restructuring would include a name change. The IFCS intends to adopt the designation "International Forum on Chemical Safety."

NGOs

Out of the large number of NGOs engaged in the field of chemicals, two of the most prominent will be featured in the following section: the International Council of Chemicals Associations (ICCA, an NGO representing business interests), and the World Wildlife Fund (WWF, an NGO representing environmental interests).

International Council of Chemical Associations

The ICCA is an umbrella organization representing 12 national or regional industrial groups. The ICCA was founded in 1989 to coordinate member associations in the area of chemical safety, develop and discuss strategies, and present their common views to IOs such as the WTO, OECD, and UNEP and to NGOs like the International Standardization Organization (ISO). The ICCA maintains a number of programmes connected with aforementioned activities. One of them is the ICCA HPV Initiative.[136]

WWF – The global conservation organization

The WWF, originally operating under the name World Wildlife Fund, is one of the largest and oldest conservation organizations in the world. It was founded in 1961, and is headquartered in Gland, Switzerland. WWF International maintains and coordinates an extensive network of national and regional organizations, for example, WWF UK, WWF Germany, WWF Centroamérica and WWF South Pacific (WWF 2008a; WWF 2008c). It is very active in the area of chemical safety. Within the framework of its Toxics Programme it cooperates with governments,

---

[135] IFCS VI, Final Report (IFCS/FORUM-VI/07w): 2ff.
[136] *Infra* 108.

IOs, and other NGOs on various issues such as the promotion of the persistent organic pollutants (POP) Convention and of the reduction of POPs in Africa (WWF 2008d). The Detox Campaign aims to reform European chemicals legislation and campaigns for the adoption of REACH, the European Commission's proposal for a new regime (WWF 2008b).

Activities

Activities in the field of international chemical safety are too numerous to describe in detail. However, the most comprehensive and substantial will be discussed here, particularly those relating to the underlying theme of this book: the interplay of legal regimes. Thus, the activities described here are those that have some sort of interaction with national legal orders. To be noted is the broad range of legal mechanisms employed, from political declarations of intent to legally binding international treaties.

Policy formulation

"Policy formulation" encompasses the setting of agendas or working plans. These are mainly non-technical and informal in nature, and provide guidance in further work and are referred to as "assignments."

The Stockholm Declaration on the Human Environment

The UNCHE of 1972 was the first "World Conference" convening representatives from 113 states in Stockholm to discuss environmental issues that had emerged in the early 1970s.[137] One outcome of this conference was a declaration that proclaimed 26 general principles.[138] Relevant to international chemical safety are Principles 6 and 7, which call for the reduction of toxic emissions. Principles 20, 22, 24, and 25 promote cooperation among states, especially with regard to scientific research exploring environmental problems. At the same time, the UNCHE adopted an Action Plan for the Human Environment, which made Recommendations for action at the international level. Recommendations 70–85 address various aspects of the hazards and risks posed by toxic substances. Of special significance is the aforementioned suggestion in Recommendation 74 (e) to establish the IRPTC.

---

[137] Cf. Birnie/Boyle 2002: 38–40 for details on UNCHE.

[138] Declaration of the United Nations Conference on the Human Environment, the Stockhold Declaration.

Chapter 19 of Agenda 21

Twenty years after UNCHE, UNCED was held in Rio de Janeiro. Representatives from 172 attended this conference. They were joined by 2400 representatives from NGOs, who had only consultative status. One of the outcomes was Agenda 21, a comprehensive plan of action to address numerous global issues. Chapter 19, headed "Environmentally Sound Management Of Toxic Chemicals, Including Prevention Of Illegal International Traffic In Toxic And Dangerous Products," sets out a framework for the taking global action to combat the most problematic aspects of chemical use. Although it recognizes the economic importance of chemicals, it also points out that the environmentally sound use of toxic substances worldwide is crucial for sustainable development. Chapter 19 identifies the two important problems connected to chemical use: first, the lack of sufficient data in this area, and second, the lack of resources to generate the necessary information for risk assessment (Chapter 19, para. 19.1). In order to realize the goal of sound chemical management, six programme areas were identified (Chapter 19, para. 19.4):

1. Expansion and acceleration of the international assessment of chemical risks (risk assessment).
2. Harmonization of classification and labelling of chemicals (classification and labelling).
3. Information exchange on toxic chemicals and chemical risks (information exchange).
4. The establishment of risk reduction programmes (risk reduction).
5. Strengthening national capabilities and capacities for management of chemicals (risk management).
6. The prevention of illegal international traffic in toxic and dangerous products (illegal traffic).

The implementation of these programme areas depends on close coordination and cooperation at the international level, and also suggests that the IPCS should serve as the nucleus for international cooperation (Chapter 19, paras 19.5f.). Chapter 19 also promoted the establishment of the IFCS, as well as the creation of the IOMC. Each programme area sets out defined objectives and recommends specific actions and means of achieving them. The programme areas picked up existing and planned measures and put them into the context of a global strategy for sustainable development. Many activities in the area of international chemical safety can be directly tied to Chapter 19.

Other parts of Agenda 21 also relate to matters of chemical safety. For example, Chapter 17, which deals with the marine environment, calls upon states "to reduce water pollution caused by organotin compounds used in anti-fouling paints."[139]

Agenda 21 is much more comprehensive than the UNCHE Action Plan, attempting to integrate economic, environmental, poverty, and development issues (Birnie and Boyle 2002: 43). Chapter 19 does not only foresee possible ways to maintain chemical safety on a global scale, but also takes into account the economic aspects and the special needs of developing countries and countries with economies in transition.

Plan of Implementation

In order to evaluate the implementation of Agenda 21 and discuss further issues of sustainable development, the World Summit on Sustainable Development (WSSD) was held in Johannesburg in 2002. The summit was attended by 10,000 governmental delegates, 100 of whom were heads of state or of government, and an additional 10,000 represented NGOs or companies.

At the summit, a Plan of Implementation was adopted. Para. 22 of the Plan of Implementation ties in with Chapter 19 of Agenda 21, and the commitment to achieve the goals set out in the six programme areas of Chapter 19 was renewed. Most notable was the goal, "to achieve by 2020 that chemicals are used and produced in ways that lead to the minimization of significant adverse effects on human health and the environment." The appropriate measures are outlined in seven subparagraphs:

- ratification and implementation of relevant international instruments, such as the Rotterdam Convention on Prior Informed Consent and the Stockholm Convention on Persistent Organic pollutants;
- development of a strategic approach to international chemicals management based on the Bahía-Declaration;
- implementation of the Globally Harmonized System for the Classification and Labelling of Chemicals;
- efforts to prevent international illegal trafficking of hazardous chemicals and wastes;
- development of coherent and integrated information on chemicals; and
- the reduction of risks posed by heavy metals.

---

[139] Para. 17.32, Chapter 17, Agenda 21.

Thus, the Plan of Implementation is not intended to reiterate the programme areas of Chapter 19, but instead reinforces and concretizes commitment to them by promoting the adoption of specific measures.

Bahía-Declaration on Chemical Safety

In 2000, the participants of IFCS III adopted the Bahía Declaration on Chemical Safety and a set of Priorities for Action Beyond 2000.[140] The Bahía Declaration reaffirms the participants' commitment to Agenda 21, Chapter 19. Article II specifies six priorities to be reviewed at future forums. Article V lists rather authoritatively the key goals of the Priorities for Action beyond 2000 (for example, "[b]y 2001: the Convention on Persistent Organic Pollutants will have been adopted..." and "[b]y Forum IV in 2003: the Rotterdam Convention will have entered into force...").

The ICCS and IFCS I had already identified Priorities for Action for each programme area defined in Chapter 19. IFCS III revised these priorities. Delineated by programme area, it clearly defines and sets deadlines for the implementation of specific actions (for example, "...through the industry initiative an additional 1000 chemicals hazard assessments will be provided by 2004...").

Strategic Approach to International Chemicals Management

The Bahía Declaration and Priorities for Action gained importance through endorsement by UNEP Governing Council, and is considered to be the foundation of the Strategic Approach to International Chemicals Management (SAICM).[141] SAICM became the main issue at Forum IV. A Thought Starter Report identifies obstacles, gaps, and omissions in the Bahía Declaration and Priorities for Action beyond 2000, and indicates potential areas where future action could be taken.[142]

The idea of a strategic approach to global chemical management is, however, much older, as one interviewee explained. "...[I]t's got its roots back in '96 when there were some countries, Stockholm...Sweden, Denmark, I can't remember if it was Norway or not, were pushing for a framework convention on chemicals. Sort of like you have the Montreal Protocol...."[143]

---

[140] The Bahía Declaration and the Priorities for Action are included in the Final Report of IFCS III in Salvador da Bahía, 20 October 2000.

[141] Governing Council Decision SS. VII/3.

[142] IFCS IV, Thought Starter Report to SAICM PrepCom1.

[143] Interview (b), 14 October 2005: 7.

The International Conference on Chemicals Management (ICCM) eventually adopted SAICM in 2006. SAICM consists of the Dubai Declaration on International Chemicals Management, the Overarching Policy Strategy, and the Global Plan of Action. All ICCM participants, not only governmental or intergovernmental actors, but also representatives from civil society and the private sector, adopted the Dubai Declaration, which acknowledges past undertakings, stating in para. 2 that "[s]ignificant, but insufficient, progress has been made in international chemical management through the implementation of chapter 19 of Agenda 21 and [ILO] Conventions No. 170…and No. 174…and the Basel Convention on the Control of Transboundary Movements of Hazardous Wastes and Their Disposal, as well as in addressing particularly hazardous chemicals through the recent entry into force of the Rotterdam Convention…and the Stockholm Convention…." The Declaration pledges to fulfil various goals concerning global chemical safety, the latter tying in with previous documents that identified aims, including: Chapter 19 of Agenda 21, and the Plan of Implementation of the Bahía Declaration (cf. para. 13). In order to achieve these goals, the declaring parties committed themselves to cooperation and solidarity (paras. 9 and 14).

The Overarching Policy Strategy consists of seven parts, specifying the scope, objective, principles, and approaches of SAICM. Part 3, entitled "Statement of Needs," identifies gaps and shortcomings in the current global management of chemicals, pointing out, *inter alia*, that the synergies between existing institutions are not fully developed, that there is both a lack of information and a insufficient information flow, and that certain countries lack the capacity to manage chemical use effectively. The fourth part establishes five objectives in the following areas: risk reduction, knowledge and information, governance, capacity building and technical cooperation, and illegal international traffic.

The Global Plan of Action ties in with the objectives of the Overarching Policy Strategy and defines several work areas. For each work area, the Global Plan of Action details activities to be carried out in order to achieve the respective objective, identifies actors, determines a timeframe, and stipulates indicators of progress.

Standardization

Another set of activities connected to the area of international chemical safety concerns matters of standardization, or harmonization. To improve the regulation of chemicals, common standards are agreed upon, for example, by increasing the comparability of test results.

Included here are, for example, OECD Test Guidelines, the UNRTDG, and the GHS.

OECD Test Guidelines

The origins of the OECD Test Guidelines Programme can be traced back to a Council Recommendation of 1977 concerning the development of methods to predict the effects of chemicals on human health and the environment.[144] The recommendation required that EPOC establish a programme to take appropriate measures. This programme was finally adopted through Council Decision C (81) 30. The decision also stipulated the principle of MAD: data which has been generated in accordance with the Test Guidelines and under observation of GLP Principles in one member state have to be accepted by other member states. This avoids the duplication of work: where companies intend to place a chemical on several national markets, they do not have to carry out multiple tests for the same chemical if they intend to place a chemical on several national markets. In addition to minimizing the industry's costs for chemical testing, this also reduces the number of test animals in the interest of animal protection, and reduces non-tariff trade barriers.[145]

Test Guidelines have also been adopted for the testing of certain physicochemical properties, effects on biotic systems, degradation and accumulation, and health effects (OECD 2007). The Test Guidelines are complemented by the GLP Principles, which cover the organizational aspects of testing (for example, organization of the facility and its personnel, as well as the apparatus, material, and reagents to be used for testing). The Test Guidelines detail the necessary testing procedures required to investigate a specific property of a substance, including exact definitions of the respective property, for example, "acute oral toxicity is the adverse effects occurring within a maximum period of 96h of an oral administration of a single dose of test substance" (OECD 1998). The testing procedure is detailed step by step. For example, in cases where animal testing is required, the Test Guidelines set out specifications regarding the husbandry of the test animals and duration of exposure.

A Guidance Document endorsed by the Joint Meeting lays down the procedure for revising existing or creating new Test Guidelines (Diderich 2007: 624ff.; OECD 2006). The procedure consists of two

---

[144] OECD Recommendation C (77) 97/Final.
[145] Interview (a) 19 July 2005: 3; cf. also Diderich 2007: 624.

phases, characterized by intense debate between participating officials. "Until they [the Test Guidelines] are finished, there is a heap of trouble,"[146] The Secretariat tries to moderate the process and considers itself a "negotiator and facilitator."[147]

The first phase consists of identifying areas where existing testing methods need to be revised or new methods need to be developed. The actual elaboration of the method happens in the second phase (Koëter 2003: 13).

The first phase begins when National Coordinators or the European Commission submit proposals to the Secretariat. The scientific community or industry can make proposals and submit these to a National Coordinator, who reviews, and, if appropriate, forwards them to the Secretariat (OECD 2006: para. 12). The Secretariat cannot reject proposals, but if it sees that crucial information is missing, it advises the submitting National Coordinator to revise his proposal before presenting it to the WNT.[148]

The WNT discusses the necessity of following the proposals (OECD 2006: paras 22ff.). If the WNT recognizes the need to create a new Test Guideline or revise an existing one it can decide that at first a Detailed Review Paper (DRP) shall be drawn up or. The DRP is normally prepared by a member country and gives detailed information on the necessity of developing or revising Test Guidelines. If the necessity is already established, the WNT can also waive the DRP and decide that a draft of the Test Guideline be prepared immediately.

The second phase includes the actual development of the Test Guideline. National Position Papers containing the views of each Member State have to be prepared by the National Coordinators. These papers ideally include concurring views of national experts on the specific issues relating to the DRP or the draft. When no scientific consensus can be achieved, alternative views may also be included. Broad scientific consensus is ultimately necessary for the worldwide acceptance of the Test Guidelines. In order to facilitate the reaching of consensus, the Secretariat may organize formal OECD Workshops, as well as ad hoc Expert Meetings on the individual aspects of a DRP or draft guideline. The Secretariat's decision to organize such events is made in consultation with the National Coordinators, and must be approved by them and the Joint Meeting. The Secretariat also calls for position

---

[146] Cf. Interview 15 December 2004: 6.
[147] Interview (a) 19 July 2005: 7.
[148] Interview (a) 19 July 2005: 4.

papers from the BIAC, the scientific community, and NGOs. The final draft of the Test Guideline must be approved by the WNT. Afterward, the Joint Meeting reviews the draft, taking into consideration the work programme and consequences for national policies, and finally decides whether to endorse the draft, or reject it and refer it back to the National Coordinators. In cases where the National Coordinators cannot achieve consensus at the technical level, the issue may be referred to the Joint Meeting. Here, the issue can be approached from a policy perspective, where it may eventually be resolved and remitted back to the WNT.[149] If the Joint Meeting endorses the draft, it will be submitted to the EPOC for an additional review. If EPOC comments on the draft, the Secretariat will either clarify the issue or refer the draft back to the National Coordinators. In practice, the actual discussions are concluded in the Joint Meeting, as EPOC usually does not remark on the drafts or otherwise participate in the process. If no feedback is received, the Secretariat will submit the draft to the Council for formal adoption. Upon adoption, the Test Guideline will become an integral part of C (81) 30/Final.[150]

UN Recommendations on the Transport of Dangerous Goods

UNCETDG developed the first internationally approved system for classification and labelling of the transport of dangerous goods, and in 1956 the UNRTDG were published.[151] The Recommendations are constantly being updated, and since 1996, have been published with an extensive Annex that details Model Regulations.[152] The Model Regulations are intended to facilitate the direct integration of the Recommendations into existing national and international regulations, enhancing harmonization.

The Model Regulations form the greater part of the Recommendations. In seven parts, which are further subdivided into several chapters, they detail provisions on various aspects of transporting dangerous goods, for example, the training of employees, packaging, and consignment of goods or transport operations. The Regulations are both detailed and authoritative. For example, Part 1, Chapter 1.1.3 stipulates which goods

---

[149] Interview (a) 19 July 2005: 12.

[150] Interview 15 December 2004: 24 and 28–9

[151] UNCETDG, UN Recommendations on the Transport of Dangerous Goods (ST/ECA/43-E/CN.2/170); endorsed by ECOSOC Resolution 645 G (XXIII).

[152] UN Recommendations on the Transport of Dangerous Good, Model Regulations, 13th revised edition (ST/SG/AC.10/1/Rev.13).

are prohibited from being transported, and Part 5 specifies requirements for every step of consignment.

The overall aim of the Recommendations and Model Regulations is to create a basic scheme to help national and international regulators establish a uniform system for the various modes of transport (Jones and Yeater 1992: 310). The revised versions of these directives are regularly endorsed by ECOSOC in a resolution requesting the UN Secretary-General to circulate and publish them and calling upon governments to consider them.[153]

Today, several international treaties or annexes to international treaties governing the transport of dangerous goods are based on the UNRTDG: the European Agreement Concerning the International Carriage of Dangerous Goods by Road,[154] the European Agreement Concerning the International Carriage of Dangerous Goods by Inland Waterways,[155] and the Regulations Concerning the International Carriage of Dangerous Goods by Rail, an annex to COTIF.[156] All contain provisions concerning the nature of substances that may be transported and the appropriate safety measures, and have their basis in the UNRTDG, so they are, therefore, closely harmonized (Jones andYeater 1992: 314). Similar rules for the transport of dangerous goods by sea and air are set out in the International Maritime Dangerous Goods Code, which is part of the SOLAS Convention,[157] and the Instructions on the safe Transport of Dangerous Substances by Air, an integral part of the ICAO Convention.[158]

Globally Harmonized System of Classification and
Labelling of Chemicals

Classifying and labelling chemicals according to their hazardous properties in order to control and communicate the risks they pose is one

---

[153]   Most recently in ECOSOC Resolution 2003/64.

[154]   *Accord européen relatif au transport international des merchandises dangereuses par route.*

[155]   *Accord européen relatif au transport international des marchandises dangereuses par voies de navigation intérieures.*

[156]   *Annexe I à l'Appendice B (Art. 4, 5 Règles uniformes concernant le contrat de transport international ferroviaire des marchandises) Règlement concernant le transport international ferroviaire des marchandises dangereuses à la Convention relative aux transports internationaux ferroviaires.*

[157]   International Maritime Dangerous Goods Code, Part A, Chapter VII of the SOLAS Convention.

[158]   International Civil Aviation Organization Technical Instructions on the Safe Transport of Dangerous Goods by Air, Annex 18 of the ICAO Convention.

of the cornerstones of chemical safety. While the relevant rules for the transport of hazardous substances have been successfully harmonized, similar regulations pertaining to other aspects of chemical safety such as environmental protection, health, and workplace safety need not exist. There are differences in the various classification systems that exist worldwide. For example, a substance must be classified as "very toxic" in the EU if the lethal dose for 50 per cent of a rat population is less than or equals 25 mg per kilogram body weight (LD_{50} 25mg/kg).[159] The United States use a cutoff value of 50mg/kg. This difference may be negligible from a toxicological perspective, but results in varying labelling requirements. Thus, the warnings and prescribed safety measures likewise differ; and since substances intended for international trade must be classified and labelled accordingly, one substance may carry several labels (Silk 2003: 447–8).

In 1989, the International Labour Conference addressed this issue,[160] drafting a resolution on workplace safety suggesting that the initial task involved harmonizing national and regional classification systems and recommending broad cooperation between other IOs and governments, as well as employers' and workers' organizations. A chemicals convention adopted by the International Labour Conference in 1991 was a further approach by the ILO towards a harmonized system[161]. However, to date, only 12 countries have ratified this convention (ILO 2008c). In 1991, the Joint Meeting adopted the ILO's suggestions and endorsed OECD participation in international harmonization activities. Other IOs joined in this endeavour. This eventually led to the creation of a Coordination Group for the Harmonization of Chemical Classification Systems (CG/HCCS) by the ILO, WHO, UNEP, UNCETDG, and OECD within the IPCS framework. Chapter 19 of Agenda 21 gave an additional impetus to press ahead with such activities, providing a comprehensive mandate (Obadia 2003: 108). Para. 19 subpara. 27 formulates the objective that "a globally harmonized hazard classification and compatible labelling system, including material safety data sheets and

---

[159] Annex VI General Classification and Labelling Requirements for Dangerous Substances and Preparations, Nr. 3.2.1., Directive 67/548; for general remarks on the European system of classification and labelling cf. Rehbinder 2003: marginal ns 115–23.

[160] International Labour Conference, Resolution concerning harmonization of systems of classification and labelling for the use of hazardous chemicals at work.

[161] Art. 6ff. of the Convention concerning Safety in the use of Chemicals at Work (C170).

easily understandable symbols, should be available, if feasible, by the year 2000." From 1995 on, the CG/HCCS operated under the umbrella of the IOMC. In 1996, the IOMC issued ToR for the work of the CG/HCCS (CG/HCCS ToR).[162] The group was further enlarged by representatives from IOs, states, and NGOs, such as WWF and the International Federation of Chemical, Energy, Mine and General Workers' Unions (ICEM).[163] As a result, the number of participants involved was rather large, since broad expertise was needed to address the complex aspects of this issue (Silk 2003: 448). A Chairperson and a Vice-Chairperson were elected from among the state representatives and the ILO provided secretarial services.[164]

In order to structure the work, the CG/HCCS devised ten general principles to guide the development and set out the scope and purpose of the GHS (Pfeil, Gerner and Vormann 2000: 306). The first and probably most important principle states: "[t]he level of protection offered to workers, consumers, the general public and the environment should not be reduced as a result of harmonizing the classification and labelling systems."[165]

Four existing classification and labelling schemes formed the basis for the development of a harmonized system: the UNRTDG, European Directives 67/548 and 99/45, the Canadian Workplace Hazardous Materials Information System, and the US Occupational Health Administration. Other systems were also taken into account, if applicable. For example, the Japanese system regarding acute toxicity was incorporated into the work (Id.: 307; Silk 2003: 448).

Harmonizing these systems turned out to be a difficult task, mainly because of diverging risk cultures. One expert contrasted European and US approaches to classification and labelling. In Europe, he explained, classification and labelling immediately follow from the hazards of a chemical: "We [Europe] have a standard test which tells us whether there is a hazard." The US system also takes into account how a chemical is used as in the end, classification and labelling depend on exposure, not merely hazardous properties.[166] Agreeing on a cutoff value for sensitizing substances exemplifies these difficulties. Canada and the United States insisted on classifying and labelling a product as sensitizing if

---

[162] IOMC, Revised Terms of Reference and Work Programme (IOMC/HCS/95).
[163] Annex 3 CG/HCCS ToR.
[164] Para. 3 CG/HCCS ToR.
[165] Chapter 1, Section 1.1.1.6, GHS.
[166] Interview 27 April 2004: 8.

it contains 0.1 per cent of a sensititzing substance to which Europe responded: "That is too much, this doesn't help anyone. We draw the line at a content of 1 per cent."[167]

The Coordination Group identified four areas of harmonization and assigned either the OECD or the ILO as respective focal points. Health hazards and dangers to the environment as well as methodological matters were the responsibility of the OECD, while the ILO attended to hazard communication, and together with the UNCETDG, physical hazards. Within each area, subgroups chaired by countries or IOs worked on specific aspects such as reactivity or carcinogenicity (ILO 2008a). The Coordination Group had the task of integrating the outcomes of the separate work areas into a systematic approach (Silk 2003: 449). The result was the GHS.

The GHS consists of two main elements: harmonized criteria for the classification of substances and mixtures according to their health, environmental, and physical hazards, and a set of harmonized hazard communication elements.[168]

Concerning the first element, the harmonized classification criteria, GHS distinguishes three broad groups of hazards:

- physicochemical hazards,
- toxicological hazards,
- ecotoxicological hazards.

Each group is subdivided into several hazard classes, which distinguish a number of hazard categories. A hazard class describes the type of hazard: for instance, explosibility is a hazard class within the category of physicochemical hazards;[169] acute toxicity is a toxicological hazard.[170]

Hazard categories reflect the severity of the hazard within a specific hazard class. For example, acute toxicity is defined as "...those adverse effects occurring following oral or dermal administration of a single dose of a substance, or multiple doses given within 24 hours, or an inhalation exposure of 4 hours."[171] Cutoff values are defined according to the exposure routes a substance takes to enter an organism. Examples for exposure routes are oral ingestion, inhalation, and

---

[167] Interview 27 April 2004: 9.
[168] Part 1, Chapter 1.1, Section 1.1.2.1 GHS.
[169] Part 2, Chapter 2.1 of GHS.
[170] Part 3, Chapter 3.1 of GHS.
[171] Part 3, Chapter 3.1, Section 3.1.1 of GHS.

dermal absorption. The respective LD_{50} or LC_{50} values determine the hazard category. Table 8.1 details the categories established for oral exposure.[172]

In some cases, the hazard category is not defined by a numerical value. Instead, the relevant factor is "weight of evidence."[173] All available information regarding toxicity is considered as a whole; it is the toxicologist's task to decide on the proper classification (Pratt 2002: 11). It should also be noted that GHS does not regulate the exact generation of data; it is neutral in this regard, allowing for different scientific approaches.[174]

The second component of GHS regulates the labelling of substances. GHS contains provisions on uniform symbols, signal words, hazard and precautionary statements, and Safety Data Sheets (SDS).[175] The symbols used are similar to those stipulated in the UNRTDG Model Regulations (for example, skull and crossbones, flame). The signal words are "danger" for more severe hazard categories, and "warning" for less severe ones. The hazard and precautionary statements are similar to the R and S clauses and SDS in Annex VI of Directive 67/548 and aim to provide comprehensive information to employers and workers about a particular chemical. GHS also lays down the required format and content of an SDS (Pratt 2002: 12ff.).

GHS does not fully harmonize the classification and labelling of chemicals. This was accepted as a compromise in certain contested areas in order to reach consensus for the overall system. Otherwise, the conflicting approaches pointed out above would not have been resolved. GHS is not a static system: maintained and promoted by the UNSCEGHS, its current inconsistencies will be worked out over time (Lowe 2003: 203–4). In fact, the industry is pushing the UNSCEGHS to

Table 8.1   GHS categories for oral exposure

Exposure route	Category 1	Category 2	Category 3	Category 4	Category 5
Oral (mg/kg)	5	5–50	300	2000	5000

---

[172] Part 3, Chapter 3.1, Table 3.1.1 of GHS.
[173] Part 1, Chapter 1.3, Section 1.3.2.4.9 of GHS.
[174] Part 1, Chapter 1.3, Section 1.3.2.4.1 of GHS.
[175] Part 1, Chapters 1.4 and 1.5 of GHS.

delete those provisions in GHS that allow for exceptions and diverging regulations.[176]

The participants of the IFCS decided that GHS should be implemented as a non-binding instrument; however, the later adoption of a binding version was not ruled out.[177] It should be noted, nevertheless, that the non-binding character allows for a certain flexibility that is necessary for the quick development and advancement of GHS. The UNRTDG are a successful example.

GHS was adopted by the newly created UNSCEGHS in December 2002. This decision was endorsed by the UNCETDG/GHS.[178] ECOSOC approved of this in a resolution by "[e]xpress[ing] its deep appreciation…," "[i]nvite[d] all Governments…" to implement GHS and requested the UN Secretary-General to publish it.[179]

GHS is a product of scientific consensus. Although it is not legally binding, it may have the function of a protolaw. States lacking a proper classification and labelling system may use GHS as a model for the creation of their own regulatory regimes; this is the case with many developing and emerging states.[180] States with an existing system are obliged to make the make the necessary changes in order to benefit from the labour taken to create GHS.

Substance reports

A number of activities can be included under the heading "substance reports." These are mainly reports on specific substances – compilations of relevant data – that reflect a consensus of international experts regarding the hazards and risks of the respective substances. Examples for such reports are the Screening Information Data Sets (SIDS) prepared in the framework of the OECD HPV Programme and the Concise International Chemical Assessment Documents (CICADs) developed under the auspices of the IPCS.

The common purpose of both programmes is the collection of data in order to close the gaps in knowledge. In this regard, they are especially useful for countries that do not command the resources to maintain their own chemicals assessment programmes.

---

[176] Interview 27 April 2004: 12.

[177] IFCS II, Final Report (IFCS/FORUM-II/97.25w), para. 26.

[178] UNCETDG/GHS, Report of the Committee of Experts on Its First Session (ST/SG/AC.10/29), para. 16.

[179] ECOSOC Resolution 2003/64.

[180] Interview 27 April 2004: 13; UNSCEGHS, Report of the Sub-Committee of Experts on Its Tenth Session (ST/SG/AC.10/C.4/20), paras 1–4.

OECD HPV Programme

In order to initiate a systematic analysis of existing substances,[181] the OECD Council decided in 1987 that OECD member states should strengthen their efforts to gather data on the properties of these chemicals systematically.[182] In 1990, the Council recognized the benefits of a concerted approach and considered that a cooperative effort of the member states would utilize the national and international resources more efficiently.[183] HPV chemicals are substances manufactured in a quantity of more than 1000 tons per year in at least one member states. These do not necessarily have to be the most hazardous substances. But as exact data on workplace, consumer or environmental exposure is obviously not always available the production volume serves as a surrogate for exposure data.[184] More than 4800 HPV-substances have so far been identified (OECD 2004b).

Originally, the objective was to collect the necessary data for all these substances and carry out a risk assessment[185]. This approach did not turn out successfully. The original programme was replaced by a refocused HPV-Programme and the work now concentrates on the compilation of data on the hazardous properties of HPV chemicals. The data is collated in SIDS. These are supposed to contain the minimum amount of information required to carry out an initial hazard assessment (OECD 2008a: Chapter 2, Section 2.1). Their content is organized under five headings: substance information, physicochemical properties, environmental fate, environmental toxicity, and mammalian toxicity (OECD 2008a: Chapter 2, Section 2.2.2), through which the most important endpoints are covered. Available data on endpoints that are usually not covered by SIDS – like sensitization or carcinogenicity – should be included if the findings were positive (OECD 2008a: Chapter 2, Section 2.4.3.3). Although it is not required information, OECD officers suggest that SIDS should also include information concerning ozone depletion or global

---

[181] Existing substances are usually chemicals manufactured or placed on the market prior to the introduction of a systematic regime to investigate the properties of the chemicals; cf. for example in the EU Art. 2(e) of Regulation 793/93; in the USA *e contrario* (reverse argument) section 3 No. 9 in conjunction with §8(b) of the TSCA; in Japan Art. 2 para. of 7 Law No. 117.

[182] OECD Decision Recommendation C (87) 90/Final.

[183] OECD Decision Recommendation C (90) 163.

[184] Annex I, Section II No. 5 OECD Decision Recommendation C (87) 90/ Final.

[185] Interview (b) 19 July 2005: 2.

warming potential of a substance – because including this information "makes sense" and "it's what the people want to see."[186]

The work begins with the selection of a chemical from the HPV list by a so-called sponsor country (OECD 2008a: Chapter 1, Section 1.3). To avoid duplication of work, OECD suggests that substances being evaluated in other programmes should not be chosen. After selecting a chemical, the sponsor country must assemble the existing data. Such data can be retrieved from companies manufacturing the substance or from databanks. The data is compiled in a preliminary SIDS Dossier. On this basis of this initial report, the sponsor country draws up a so-called SIDS Plan, which determines the necessity of generating new data through additional tests. In Electronic Discussion Groups (EDGs), member state representatives discuss the SIDS Dossier and decide on the adoption of the SIDS Plan. Tests have to be carried out according to OECD Test Guidelines and GLP Principles (OECD 2008a: Chapter 2, Section 2.3), and the SIDS Dossier is amended accordingly.

On the basis of the completed SIDS Dossier, which provides the relevant information for a hazard assessment, the sponsor country prepares a SIDS Initial Assessment Report (SIAR), a summary and evaluation of available data and initial assessment of the hazard posed by the chemical to human health and the environment. A recommendation is made on this basis either that "the chemical is currently of low priority for further work" or "the chemical is a candidate for further work" (OECD 2008a: Chapter 5, Section 5.2.2). Companies are very keen to see their substance receive a low-priority recommendation. One interviewee observed: "They can go to their government and say 'See, all the work has been done....'".[187]

A SIDS Initial Assessment Profile (SIAP) accompanies the SIAR, providing a short, general summary of the SIAR's content (OECD 2008a: Chapter 5, Section 5.3.2).

Both the SIAR and SIAP are discussed at a SIDS Initial Assessment Meeting (SIAM). This meeting is attended by representatives from sponsor countries, other member states and the EC, experts from non-member states nominated by the IPCS and IFCS, BIAC, TUAC, and NGOs, representatives from companies that produce the screened substance, and secretarial staff from the OECD, IPCS, and UNEP Chemicals. Participants must reach a consensual agreement on the assessment of the chemical (OECD 2004a).

---

[186] Interview (b) 19 July 2005: 13.
[187] Interview (b) 19 July 2005: 19.

The sponsor country then finalizes the SIAR and SIAP, taking into account the comments made at the SIAM, and submits them, along with the SIDS Dossier, to the OECD Secretariat. The Task Force on Existing Chemicals oversees the whole process and confirms the conclusions and recommendations agreed upon at the SIAM, which are then submitted to the Joint Meeting for endorsement (OECD 2008a: Chapter 1, Section 1.4). The SIDS Dossier and SIAR are made publicly available through UNEP Chemicals.[188]

Subsequent work is described as Post-SIDS-Activity (OECD 2008a: Chapter 6). This includes national efforts to collect and assess test data. Since the OECD focuses almost entirely on the preparation of SIDS, not much Post-SIDS-Activity is carried out within this framework. Instead, member states implement measures they deem necessary and appropriate.

OECD's organizational role is to provide secretarial services and comment on the formal aspects of the process with the aim of improving the programme and ensuring its coherence. Occasionally, OECD officers will also get involved in the discussions on the content of the SIDS. The actual task of compiling, assessing, and discussing the data and assembling the SIDS is carried out by the competent authorities of participating states.[189]

In 1998, US launched its own HPV Initiative. The background to this industry-managed programme is the US HPV challenge. Following the Environmental Defense Funds' (EDF) "Toxic Ignorance" Study,[190] the EPA and then Vice-President Al Gore called upon the US chemicals industry to close this gap in knowledge voluntarily by compiling data in accordance with OECD SIDS Programme (EPA 1998b). In order to restore public confidence in the chemicals industry, avoid the duplication of work, save on costs, reduce the number of test animals used, and produce internationally harmonized and coherent data sets, the ICCA internationalized the effort by launching the ICCA HPV Initiative (Kistenbrügger 2004: 38; CEFIC 2008d).

Although the HPV Initiative is a separate endeavour, it is closely linked to the OECD HPV Programme. ICCA has compiled a Working List of more than 1300 substances using the OECD List of HPV Chemicals (ICCA 2005). Companies have been invited to collect and generate data and prepare SIDS, SIAR, and SIDS Profiles for these substances (CEFIC

---

[188] SIDS for about 400 substances are available at UNEP 2008.
[189] Interview (b) 19 July 2005: 4–5 and 13.
[190] *Supra* 74.

2008d). To facilitate the work, companies form consortia to share the costs of the investigation (CEFIC 2008c).[191]

After compiling and collating the necessary information in accordance with the SIDS formula, the companies or consortia submit the SIDS to a sponsor country. Thus, the SIDS prepared under the ICCA HPV Initiative feed into OECD HPV Programme. The sponsor country reviews the SIDS and then submits it to OECD, where it will be regularly reviewed in a SIAM (Kistenbrügger 2004: 39).

Although active participation of the chemicals industry is necessary to generate the data to complete SIDS, one major problem hampers the success of the HPV Initiative. Companies may be willing to cooperate in the global public interest and join such an endeavour. Nevertheless, ultimately companies compete and are not inclined towards sharing presumably sensitive data concerning the substance of properties with their competitors or engaging in costly effort that does not increase company earnings. Consequently, only a fraction of the substances from the Working List – ca. 400 from a list of more than 1000 – has been investigated (Winter 2007: 826–7).

The HPV Programme clearly follows a "ton-philosophy." According to this approach, the thoroughness of the investigation depends on the manufactured quantity of a substance (Warning and Winter 2003: 255). According to one expert, up to now, the investigation of HPV substances has not revealed any surprises; in most cases, hazardous properties were already known or suspected.[192] It is on this basis the quantity-oriented approach can be criticized. Chemicals manufactured in lower quantities could pose even greater risks that might ultimately be overlooked. Furthermore, the concentration on production volume means that so-called intermediate substances are also evaluated. However, intermediate substances are usually used only in the production process; this means that consumers and the environment scarcely come into contact with such substances, reducing their actual risk.

Differences in the risk cultures of participating states can hamper the efficacy of the programme. For example, pursuant to Nos. 28–30 of Annex XVII of Regulation (EC) No. 1907/2006, the same legal consequences arise with the classification of a substance as carcinogenic, mutagenic, or toxic to reproduction.[193] If a substance has one of these properties, it must not be used in substances and preparations placed

---

191 List of consortia available at ICCA 2008.

192 Interview (b) 19 July 2005: 26–7.

193 Until recently, this was regulated in Nos. 29–31 of Annex I of Council Directive 76/769/EEC. According to Art. 139 of Regulation 1907/2006, Council Directive 76/769/EEC was repealed with effect from 1 July 2009.

on the market for sale to the general public in excess of certain concentrations and the packaging of such substances and preparations must be marked legibly and indelibly with the phrase "Restricted to professional users." As a result, European countries are less inclined to discuss the other properties of a chemical in instances where it has already been determined to fall within one of these categories. From their point of view, it is not necessary to test the substance further, because its legal treatment has already been determined. In contrast, US authorities always require data on reprotoxicity. Such differences can bring the whole process for one substance to a "cul-de-sac," as one interviewee remarked.[194]

Nevertheless, the HPV Programme aims to make as much data publicly available as possible. The amount of data is continually growing, reducing the overall gap in knowledge. Since the manufacture of HPV substances is likely to shift to non-OECD countries in the coming decades (OECD 2001: 36–7), it will be helpful for these countries to have access to a basic data set and not be burdened with the investigation of the substances' properties.

CICADs

The IPCS maintains a number of publications on chemical safety,[195] including CICADs. CICADs are short summaries of the relevant scientific data on the possible effects of a chemical on human health and/or the environment. Usually, they are based on an existing document that is converted into a specific format and amended according to comments made in the peer review process. CICADs contain critical information not only on a chemical's hazardous properties, but also on dose-response from exposure and a sample risk characterization (IPCS 2002: 1), meaning that they are broader in scope than SIDS documents.

The selection of chemicals for the preparation of CICADs is an iterative process that begins with a proposal supported by both national and international institutions. These institutions provide the resources for the preparation of the CICADs. Ideally, the institution making the proposal will later "sponsor" the CICADs.

The IPCS Risk Assessment Steering Group counsels the IPCS Director in his determination of whether a substance should be placed on the IPCS agenda, if a CICAD should be prepared, and on which institutions

---

[194] Interview (b) 19 July 2005: 6–7.
[195] For an overview cf. Hildebrandt and Schlottmann 1998: 1384–5.; Gärtner, Küllmer, and Schlottmann 2003: 4604.

or groups should prepare the draft report and which should peer review it (IPCS Risk Assessment Steering Group 2008). Substances with a high likelihood of exposure or a significant toxicity or ecotoxicity are given priority treatment.

A first draft CICAD is prepared by the sponsor, either a governmental or a private institution.[196] The draft CICAD is based on an existing national, regional, or international document. If the original document is more than two years old, the authors must conduct literature surveys and update it accordingly. After a primary review by the IPCS, which simply ensures that the specific criteria have been adhered to, the draft is subjected to an international peer review process. The peer reviewers can be scientists from IPCS Contact Points, Participating Institutions, or NGOs. They comment on the draft by submitting papers or reports considered for inclusion in the CICAD. The sponsor then prepares a second draft that takes into account the peer reviewers' comments.

Next, the second draft is submitted to a Final Review Board (FRB), whose members are selected by the IPCS Secretariat. They serve as experts, and are not representatives of any government, organization, or company. The task of the FRB is to ensure the proper conduct of the peer review, verify that the peer reviewers' comments have been appropriately addressed, provide guidance on how to resolve remaining issues, and approve the CICAD as an international assessment. If a consensus cannot be reached, the dissenting participants will be listed in the CICAD.

Representatives of the most important manufacturers of the substance are present during the draft discussions. As observers, they are not allowed to interfere with the FRB, and may only provide factual information on the substances that fall within their area of expertise. After the FRB endorses the draft, the WHO publishes the CICAD on behalf of the IPCS.[197] The protocols of the session are also made available to the public (Cf. IPCS 2008b).

All participants in the process are required to submit a Declaration of Interests (IPCS 2002: 22). Thus, commentators have to reveal potential conflicts of interests. Authors of CICADs are especially cautious with comments from industry representatives. "Of course you read their comments, but you know, you have to approach them with caution," an

---

[196] In Germany, the Fraunhofer Institute of Toxicology and Aerosol Research, based in Hanover, is usually tasked with composing a draft.

[197] CICADs for about 70 substances are available IPCS 2008a.

author explained.[198] Government officials may be influenced by their government's position, but they act first and foremost as scientists. One participant explained: "The only task of [my employer] is to pay me, as I do not work on CICADs as a representative of [my employer], but as a scientist." But he recognized that the two aspects cannot easily be separated.[199] This emphasis on a scientific basis sets CICADs apart from OECD HPV Programme, under which companies essentially write the reports. Another expert stressed the unbiased character of CICADs: "I feel safer with a document when I know that this person [who wrote it] was not paid by selling this chemical." In his opinion, this measure of objectivity adds to the CICADs' persuasiveness.[200]

Unlike OECD's HPV Programme, the selection of chemicals for CICADS does not depend on production volume. Since CICADs are based on existing risk assessments, other factors such as toxicity can be taken into account here. The IPCS is not confined to developed nations, which means that CICADs may be also be prepared for substances that affect developing countries or countries with economies in transition.

The IPCS Secretariat has addressed the issue of duplication of work with a "Rule of Thumb" that stresses the different approaches and goals of the two programmes (IPCS 2008c). In order to ensure an equal sharing of the load, OECD SIARs are not to be prepared if there is an existing CICAD which has been updated in the past five years, or if a CICAD is currently underway or planned for completion in the next two years. If a substance has been selected for a SIDS or has been in the information gathering stage for more than five years and a national assessment report is available, this should be brought to the attention of the IPCS and the OECD. They will jointly decide on further proceedings.

Information exchange

Generally, it is a state's responsibility to protect both its citizens and environment from the harmful effects of industrial chemicals or pesticides. Developing countries and countries with economies in transition often lack the resources and capacity to perform this task effectively, with disastrous results.

One option for addressing this specific aspect of chemical safety is to regulate international trade in toxic substances with such nations

---

[198] Interview 13 July 2005: 17.
[199] Interview 11 May 2004: 5.
[200] Interview (a) 14 October 2005; 19–20.

through the introduction of a prior informed consent (PIC) procedures.[201] This would allow developing states to decide which substances
might be imported and the importation of chemicals or pesticides that
cannot be safely managed might therefore be interdicted (VanDorn
1998: 281). With the exception of unilateral export licenses, the PIC
procedure will respect the sovereignty of the importing country in
instances where developed nations decide on which substances may be
exported from their territory to other countries.

Voluntary PIC procedures were introduced in 1989 with amendments
to the FAO International Code of Conduct on the Distribution and
Use of Pesticides and UNEP London Guidelines for the Exchange of
Information on Chemicals in International Trade. Neither instrument
initially included PIC procedures; rather, trade in pesticides and chemicals was to be regulated by means of information exchange and export
notification, as neither required the consent of the importing country.
In practice, these measures proved to be insufficient. Thus, revisions
were implemented in 1989, introducing a PIC procedure to both instruments (Mekouar 2000: 147–60; Ross 1999: 512–18; Pallemaerts 2003:
441–556).

The voluntary PIC procedures did not operate flawlessly. They failed
to promote the development of infrastructure and information systems,
especially in developing countries. Furthermore, as voluntary mechanisms, they were not coupled with enforcement mechanisms, so the
compliance rate was rather low (Ross 1999: 515ff.). To remedy these
shortcomings, a legally binding PIC procedure came under consideration. Chapter 19 of Agenda 21 picked up the discussion and proposed a
legally binding PIC procedure in Programme Area C.[202] Pursuant to this
suggestion, the FAO Conference and UNEP's Governing Council initiated the creation of an intergovernmental negotiating committee (INC).
The INC prepared a draft, which was finally adopted in Rotterdam in
1998 (Pallemaerts 2003: 511–72). The Rotterdam Convention on the
Prior Informed Consent Procedure for Certain Hazardous Chemicals
and Pesticides in International Trade (PIC Convention) entered into
force on 24 February 2001. It was modelled on the Code of Conduct
and London Guidelines (Mekouar 2000: 162).

The following section will discuss the most important aspects of the
PIC Convention. The main objectives of the PIC Convention, declared
in Art. 1, are the protection of human health and the environment and

---

[201] For a discussion of other options cf. Ross 1999: 508ff.
[202] Chapter 19, paras 19.38 (b), 19.39 (d), 19.52 (f), Agenda 21.

to foster global responsibility and cooperation in this area. It introduces the PIC procedure as a way of ensuring that its implementation goals will be achieved. This procedure applies to chemicals listed in Annex III of the Convention. Any party that has already taken legislative measures to ban or restrict a chemical may propose the inclusion of this substance in Annex III, under Art. 5. Likewise, according to Art. 6, any developing country or country with an economy in transition that is experiencing problems with a severely hazardous pesticide formulation may make a similar proposal. A Chemical Review Committee – a special body of government-designated experts appointed by the Conference of the Parties (Art. 18 (6)) – will review the information provided with the submitted proposal and decide whether or not to recommend the substance for inclusion in Annex III (Arts 5 (6), 6 (5)). If it recommends a substance for inclusion, Art. 7 (1) requires that the Committee draw up a draft decision guidance document, which will be submitted along with the recommendation to the CoP. Clear standards to guide the decision-making process of the Chemical Review Committee are not included in the provisions (Ross 1999: 520–1). Finally, the CoP decides whether the substance will be listed in Annex III, making it subject to the PIC procedure (Art. 7 (2)). The decision to amend Annex III must be taken in consensus (cf. Art. 22 (5) (b)). Art. 9 lays down the procedure for the removal of substances from Annex III.

Once a substance is listed in Annex III, the parties have to meet the import requirements stipulated in Art. 10. Each party is required to implement appropriate legislative or administrative measures and send a response on the future importation of the chemical to the Secretariat. The response shall either be a final decision or an interim response. A party's final decision may consent or refuse consent to importation or provide consent under specified conditions. Thus, restrictions do not constitute a total ban: a country can decide on whether or not to ban or restrict a chemical, depending on its ability and resources to manage the use of the substance safely. If a party refuses consent to importation or places specific conditions thereon, it must simultaneously prohibit or make subject to the same conditions the importation of the chemical from any other source and the domestic production. This shall prevent the misuse of the convention to introduce non-tariff barriers of trade (VanDorn 1998: 287–8).

Art. 11 contains obligations relating to the export of substances listed in Annex III. Exporting parties have to implement appropriate legislative measures to ensure that exporters within their jurisdiction comply with the decisions set out in the importing party's response. It is their responsibility to ensure that chemicals included in Annex III are not exported from their territory to any other party, unless previous imports of the substance have not been subject of regulatory measures or the exporter has sought and receive the consent of the competent authority in the importing country. Export notifications have to be issued in accordance with the requirements outlined in Art. 12. If chemicals that have been banned or severely restricted by one party, but not yet listed in Annex III, are exported from its territory, the exporting country has provide notification to the importing country. The notification has to include information concerning the substance's properties, particularly information on safety precautions (Annex V). However, although notification is mandatory, the provision does not introduce a mechanism for non-compliance (Zahedi 1999: 722; Ross 1999: 522).

Art. 17 calls upon the CoP to introduce procedures and institutional mechanisms to determine whether there has been non-compliance, and, if so, how the violating party should be treated. Possible sanctions include: punitive measures such as fines, coercive mechanisms such as the prohibition of exports of Annex III substances to countries that failed to register their consent or non-consent, or "sunshine methods," such as publishing annual reports (Zahedi 1999: 723). The Open-ended Working Group, convened by the CoP, developed plans for the establishment of a Compliance Committee. On its third meeting in October 2007, the CoP reached consensus only on a draft text for the establishment of a Compliance Committee as the basis for further work. The final decision, however, has been postponed. At its fourth meeting in October 2008, the Conference again decided on a draft text as a further basis for work. Current plans for the Compliance Committee envisage that it would be furnished with competences to respond to compliance issues, ranging from furthering support for the non-compliant member and offering assistance, to the non-compliant party's ineligibility to serve as the President of the CoP or as a member of the Bureau until it has fulfilled its obligations.[203]

According to Art. 13, exported chemicals are to be adequately labelled, providing information on hazards or risks to human health or the environment. Art. 14 calls on parties to facilitate the exchange of scientific information.

---

[203] CoP of the PIC Convention Decision RC 1-10; CoP of the PIC Convention Decision RC 3/4; CoP of the PIC Convention Decision RC-4/7.

Developing countries and countries with economies in transition are of special concern. Art. 16 takes into account their needs, calling on parties with more advanced regulatory programmes to provide technical assistance in these areas. However, the PIC Convention does not detail the exact nature of such assistance. A key aim of the PIC Convention is to help countries – particularly developing countries or countries with an economy in transition – to make informed decisions about whether or not to allow certain chemicals into their territory (VanDorn 1998: 290). The technical assistance provided to more vulnerable countries may help to build the necessary capacities ultimately to raise the standard of chemical safety. However, the dissemination of information under the PIC Convention is of only limited use if the authorities in the importing countries lack the capacity to process and assess the information properly (Zahedi 1999: 730). Unfortunately, the PIC Convention is rather vague in this respect. Furthermore, in contrast to the Convention on Biological Diversity, the Montreal Protocol, and the United Nations Framework Convention on Climate Change, mechanisms to ensure financial assistance have not been included. Options for financial measures were discussed at a session of the Intergovernmental Negotation Committee (INC), but the issue was not revisited at later meetings (Id.: 731–2; Pallemaerts 2003: 584).[204] The CoP has adopted a decision concerning this matter, which set outs measures focusing on the proper implementation of the Convention (i.e. the creation of legal infrastructure) and the Secretariat has been instructed to explore cooperation with other programs.[205] In one step towards cooperation and coordination with the POP Convention and the Basel Convention on the Control of Transboundary Movements of Hazardous Wastes and their Disposal, the Conferences of the Parties was the creation of the Ad Hoc Joint Working Group.[206]

NGOs, together with UN Specialized Agencies and other non-parties, have been admitted as observers to the CoP (Art. 18 (7)). While IOs and non-parties can participate in the proceedings of any meeting without the right to vote, NGOs can participate only in those that directly concern them (cf. Rule 6f. RoP PIC-CoP); beyond this, they play no prominent role.

---

[204] INC for the PIC Convention, 56.

[205] CoP of the PIC Convention Decision RC-1/14.

[206] Cf. Ad Hoc Joint Working Group, Report of the ad hoc joint working group on enhancing cooperation and coordination among the Basel, Rotterdam and Stockholm conventions on the work of its first meeting.

In view of its weaknesses and shortcomings, international lawyers are ambivalent in their assessment of the prospects of the PIC Convention. Its language is considered to be too vague in some respects, allowing parties to interpret the text to their advantage (Ross 1999: 520). Formally incorporating the PIC procedure into a convention, which is binding under international law, has firmed up the process; but at the same time, substantive measures have been blunted. Hence, the Convention is considered to be a form of "hard" soft law or "soft" hard law (Pallemaerts 2003: 594). Nevertheless, there are others who view it as a positive step in global chemical management (Ross 1999: 525). Some criticisms will be addressed by the CoP, since it ultimately falls on them to make the PIC Convention an effective tool.

Restrictions and bans

Restrictions and bans of certain chemicals are the most severe kind of regulatory action. International measures imposing substance restrictions or bans are rare.

Stockholm Convention on Persistent Organic Pollutants

Agenda 21 recognizes the problematic nature of POPs, but does not explicitly call for a treaty to tackle it, only demanding that governments together with IOs and industry members "undertake concerted activities to reduce risks for toxic chemicals, taking into account the entire life cycle of the chemicals" including the use of limitations such as the phasing out or banning of chemicals. Furthermore, they should "adopt policies and regulatory and non-regulatory measures to identify and minimize exposure to toxic chemicals by…ultimately phasing out…those that are toxic, persistent, and bio-accumulative and whose use cannot be adequately controlled".[207] Chapter 17, which is dedicated to the protection of the oceans, calls on states to consider actions to regulate the emission or discharge of organic pollutants into the marine environment.[208]

Pursuant to the demands of Agenda 21, UNEP's Governing Council invited the IOMC to work with the IPCS and IFCS to initiate an assessment process that should, *inter alia*, assess possible means to counter the problem. The assessment was to be initially confined to 12 substances

---

[207] Chapter 19, para. 19.49 (b) and (c), Agenda 21.
[208] Chapter 17, para. 17.28 (d)-(g), Agenda 21.

identified in the decision: PCBs, dioxins and furans, aldrin, dieldrin, DDT, endrin, chlordane, HCB, mirex, toxaphene and heptachlor.

The IPCS was also invited to participate in the process and to prepare recommendations and information regarding an appropriate legal mechanism to regulate such substances.[209] By the end of 1995, IPCS had published the requested assessment.[210] It analysed the scientific data available for the 12 substances and would later provide the scientific groundwork for future proceedings. In response to the Governing Council's decision, in 1996, the IFCS convened a meeting of experts in Manila. These experts came from developed and developing countries, as well as countries with transitional economies, and various NGOs (for example the ICCA, Greenpeace, the WWF) and IOs (for example the WHO, FAO, ILO) also participated. Together, they agreed on the scientific underpinning for action and recommended that the focus of future regulatory action remain on POPs, rather than expanding their mandate to include other chemicals.[211] In 1997, the Governing Council endorsed this process, initiating negotiations on a treaty regulating POPs. It also laid down several guiding principles for the negotiation process, which established the cornerstones of the future treaty:

- the instrument should initially focus on the 12 substances previously identified;
- it should include differentiated approaches for pesticides, industrial chemicals, unintentional byproducts, and contaminants, regulate transition periods, include provisions on the management of existing stocks, capacity building measures, and remediation of contaminated sites;
- socioeconomic factors should be considered; and,
- science-based criteria for the identification of additional POPs should be developed.[212]

The limitation to these 12 POPs was considered important in order to allow the expedition of the process. Consensus on these substances

---

[209] UNEP Governing Council Decision 18/32; for a detailed account of the drafting process cf. Olsen 2003: 77–105.

[210] IPCS, Persistent Organic Pollutants (PCS/95.38).

[211] Final Report of the IFCS ad hoc Working Group on Persistent Organic Pollutants.

[212] UNEP Governing Council Decision 19/13C.

already existed at an international level, so by concentrating on these, no momentum would be lost on tedious discussions about include further chemicals or pesticides (Lallas 2001: 696).

Treaty negotiations started in 1998. The drafting process of the POP Convention was marked by a high degree of transparency. Contributions from UNEP and other IOs moved the treaty process forward (Id.: 707–8). Although they were only admitted as observers with no negotiating role,[213] NGOs like Greenpeace International also influenced the treaty negotiations.[214] The final text was adopted in Stockholm in 2001, and the Stockholm Convention on Persistent Organic Pollutants (POP Convention) entered into force in 2004.

The following will provide a brief overview on key provisions of the POP Convention (Olsen 2003: 107–125). Art. 1 defines the objectives of the POP Convention. Considering the precautionary principle as defined in Principle 15 of the Rio Declaration, human health and the environment shall be protected from POPs. According to Art. 3, states must take the necessary legal or administrative measures to eliminate and restrict releases from intentional production and use. This includes special requirements concerning the export and import of POPs, impacting states that are not party to the Convention.[215] Annex A sets out the substances whose use or production is to be eliminated; and, Annex B lists restricted substances. Because some countries rely on certain substances for specific purposes, parties may register for a temporary exemption in accordance with Art. 4. The Annexes lay down possible exemptions for each substance. For example, Mirex may still be used as a termiticide, and DDT for disease vector control. No exemption may be made for Endrin. Parties must also take measures to reduce the release of POPs listed in Annex C from unintentional production (Art. 5). Measures include the promotion of best available techniques and best environmental practices. Finally, Art. 6 addresses releases from stockpiles and wastes.

As stated previously, negotiations focused on 12 substances because the necessity to regulate these chemicals was undisputed. In order to keep the convention dynamic and make it possible to further regulate other POPs, Art. 8 introduces a mechanism which allows parties

---

[213] Cf. Rule 55 of the RoP for the meetings of the INC (UNEP/POPS/INC.1/2).

[214] Interview 8 December 2004: 22.

[215] Cf. Art. 3 para. 2 (b) (iii). States may only export POPs to states not parties to the convention if the latter warrant their compliance with certain requirements.

to propose substances for listing in Annexes A, B, and C.[216] Proposals must meet the criteria laid down in Annex D. If the proposal fulfils the criteria, the Persistent Organic Pollutants Review Committee prepares a draft risk profile in accordance with Annex E. The Committee consists of experts sent by the governments and appointed by the CoP according to fair geographic distribution (Olsen 2003: 113). The draft is submitted to the parties and observers for comments, which are taken into account by the Committee upon completion of the risk profile. Based on the outcome of the risk profile, the Committee may decide to proceed, requesting that the parties and observers submit information relating to the socioeconomic aspects specified in Annex F. If scientific data is lacking, the proposal may still be advanced, so long as there is reliance on the precautionary principle as a justification. The Committee will prepare a risk management evaluation including an analysis of possible control measures. It may also decide that the proposal shall not proceed. On the basis of the risk profile and risk management analysis, the Committee then recommends whether the substance should be considered for listing in Annexes A, B, and/or C. Finally, the CoP decides on the listing of the chemical and the appropriate control measures, taking into account the recommendations of the Committee. Decisions regarding amendments should be unanimous. If a consensus cannot be achieved, amendments can also be passed by a three-fourths majority of those present at the CoP (cf. Art. 22 (3) (a), in conjunction with Art. 2 (3)). Art. 22 (4) allows the parties to opt out upon notifying the Secretariat that the amendment will not enter into force in their own state.

Arts 9–11 contain provisions addressing specific aspects of the overall objective of eliminating or reducing the release of POPs and prescribe "softer" measures. These include exchange of information regarding the reduction or elimination of POPs and possible substitutes; raising awareness among government officials and the public; and the undertaking of further research, development, and monitoring pertaining to POPs, their alternatives, and candidates for future inclusion in the treaty.

Arts 12 (3)–(5) and 13 (2)–(7) direct developed parties to provide technical and financial assistance to developing parties and parties with economies in transition. In addition, UNIDO maintains specific

---

[216] As of November 2008, the parties to the POP Convention had submitted 11 substances to the chemical review process, cf. Stockholm Convention on Persistent Organic Pollutants 2008a.

programmes focusing on capacity building in the area of POP elimination (UNIDO 2008a). There are also initiatives taken at a local level, which try to raise awareness of POPs, and the promotion of notions of Best Available Techniques and Best Environmental Practices. Since POPs are often produced and used in developing countries or countries with transitional economies, the assistance mechanisms are crucial to the success of the POP Convention. Without assistance or incentives, these countries are not likely to discontinue the production and use of such chemicals.

The POP Convention does not currently contain provisions dealing with non-compliance. According to Art. 17, appropriate procedures and mechanisms are to be developed and approved by the CoP. In a draft text, which will form the basis for further negotiations on the issue of non-compliance, the CoP considers measures in response to non-compliance like providing further assistance to the non-compliant party or even suspending rights and privileges under the Convention.[217]

NGOs are admitted as observers to the CoP, in accordance with Art. 19 (8) (2) They are allowed to participate in those meetings that directly concern them, but do not have the right to vote in the proceedings (cf. Rule 7 (7) RoP of the CoP of the POP Convention[218]). As observers, they participate in the amendment process and are invited to submit comments on the Committee's decisions. Furthermore, Art. 19 (5) (b) instructs the CoP to "[c]ooperate, where appropriate, with competent international organizations and intergovernmental and non-governmental bodies." So far, this general provision has not been translated into concrete measures.

Many observers criticise the Convention's weak and vague language as its major shortcoming. However, such broad and cautious wording is necessary when dealing with complex issues, where the conflicting interests of developed countries and developing countries collide. As a result, it is up to the parties and other actors to put the POP Convention to good use. It is unclear how the treaty will be enforced, as it does not provide guidance on implementation (Olsen 2003: 123). However, it must be noted that UNEP Chemicals now offers several documents that provide guidance on various issues concerning the implementation of the Convention (Stockholm Convention on Persistent Organic Pollutants 2008b).

---

[217] CoP of the POP Convention Decision SC-3/20.
[218] Rules of Procedure for the CoP (UNEP/POPS/COP.1/25).

Other treaties

There are at least two other international legal agreements that aim to restrict or ban certain chemicals. One is the well-known Montreal Protocol on Substances that Deplete the Ozone Layer. The Montreal Protocol was adopted in 1987 within the framework of the Vienna Convention for the Protection of the Ozone Layer ("Ozone Convention"). It regulates the phasing out of certain substances that were found to deplete the ozone layer, such as CFCs and halons. As the Ozone Convention and the Montreal Protocol have already been subject to extensive analysis, only a few core features will be mentioned here.

The Ozone Convention only establishes a framework for further measures. It contains provisions promoting further research and international legal, technical, and scientific cooperation (Art. 3f.). The introduction of specific measures is left to future protocols: the Ozone Convention only details the process of adopting and amending annexes and protocols (Arts 9–10), including the possibility of amending protocols with a two-thirds majority of the parties, if a consensus cannot be achieved (Art. 9 (4)).

The objective of the Montreal Protocol is the reduction and eventual elimination of certain ozone-depleting substances (Birnie and Boyle 2002: 519). Art. 2 sets out detailed measures that countries must undertake for each substance or group of substances. Art. 2 (9) contains a procedure for adjusting the regime. If a consensual agreement on this procedure cannot be reached, a two-thirds majority may decide to amend the adjusting measure, binding all parties (cf. Art. 2 (9) (c) (1) and (d)).

Art. 5 recognizes the special situation faced by developing countries. They may delay their compliance for ten years, if they meet certain requirements and if the annual per-capita-consumption of zzone-depleting substances does not exceed a specific level. In addition, Art. 10 establishes a financial mechanism to support implementation in these countries, and Art. 10a instructs parties to provide technical support.

In order to ensure the effective implementation of its goals and address the issue of free riders, Art. 4 directs parties to ban the trade of these substances with countries not party to the Montreal Protocol. Importation and exportation of certain substances, which are listed in Annex I, are prohibited. The Protocol allows NGOs to be admitted at the CoP (Art. 11 (4) (2), cf. also Art. 6 (5) (2) of the Vienna Convention). However, apart from this provision, the Montreal Protocol does not mention non-state actors.

The Montreal Protocol has proven to be a successful instrument. More and more countries have become party to the agreement, the controls on substances have been strengthened through successive amendments, and the compliance level has been high. However, one problem remains: substances and technologies used to substitute the ozone-depleting chemicals are often classified as greenhouse gases, creating a problem for the global climate. In this regard, there has been little coordination with the climate change regime established with the United Nations Framework Convention on Climate Change and the Kyoto Protocol (Birnie and Boyle 2002: 521ff.).

A further example of the ban or restriction of substances at the global level is the International Convention on the Control of Harmful Anti-Fouling Systems on Ships (the Anti-Fouling Convention). To tackle the problem of organotin compounds, which affect the endocrine systems of marine life, a convention was prepared within the framework of the IMO. Since a main task of the IMO is the prevention and reduction of marine pollution, it has tackled the issue of organotin compounds that are used to protect ship hulls from growth of barnacles, algae, etc. since 1990.[219] Its Marine Environment Protection Committee (MEPC) adopted a resolution recommending that countries take appropriate measures to eliminate the use of the organotin compound tributyltin in anti-fouling systems.[220] After an extensive evaluation of existing national measures, in 1998 the MEPC initiated the creation of a legally binding instrument at the international level. The Anti-Fouling Convention was finally adopted at a diplomatic conference in 2001 (Champ 2000: 33ff.). It came into force in September 2008.

The key provision of the Anti-Fouling Convention is Art. 4, which instructs parties to prohibit or restrict "...the application, re-application, or use of harmful anti-fouling systems on ships...." Annex I details the anti-fouling systems covered by the Convention. Currently, only organotin compounds are listed, but Annex 1 can be amended in accordance with the procedure laid down in Art. 6 to include other substances, too. Any party may submit a proposal to the IMO, and Annex 2 sets out the content of such proposals. The MEPC decides whether the antifouling system requires further review. If further review is required, the MEPC will request the proposing party to submit a comprehensive proposal,

---

[219] Cf. Art. 1(a) of the IMO Convention.

[220] Marine Environment Protection Committee, Resolution on Measures to Control Potential Adverse Impacts Associated with Use of Tributyl Tin Compounds in Anti-Fouling Paints (MEPC 46 (30)).

as required in Annex 3. Pursuant to Art. 7, the MEPC will establish a Technical Group composed of party representatives, the IMO, other IOs, and NGOs (cf. Art. 2 (10)), which will review the proposal and any additional data provided to determine if the antifouling system in question has the potential for the unreasonable risk of adverse effects on non-target organisms or human health. The Technical Group's decision should be unanimous. If this cannot be achieved, minority views will be communicated. In its report to the MEPC, the Technical Group is to make recommendations pertaining to the necessity of international controls, the suitability of specific measures suggested in the comprehensive proposal, and other future measures. The report is made available to the Parties, UN members, members of its Specialized Agencies, IOs, and NGOs. Afterwards, the Committee decides whether to approve the proposal to amend Annex I, taking the report into account.

Once a proposal has been accepted, amendments to Annex I can be made by two procedures. The first, set out in Art. 16 (2), involves the MEPC, which considers the proposal and can adopt it with a two-thirds majority. Upon its adoption, the amendment is referred to the parties to the Convention for acceptance. If more than one-third of parties object, the amendment is considered to have been rejected (Art. 16 (2) (e) (ii)). Moreover, the amendment does not enter into force for those countries that submit objections (Art. 16 (2) (f) (ii)). Amendments may also be adopted by a second procedure, under Art. 16 (3), which allows any party to request that the IMO convene a conference. This request must have at least one-third party support if it is to be effectual. The remaining procedure regarding the adoption and acceptance of amendments follows the same MEPC procedure.

Furthermore, the Anti-Fouling Convention obliges the parties to implement specific control measures to ensure its efficacy. Arts 10f. requires that parties inspect ships entering their ports or offshore terminals regarding compliance with the provisions of the Anti-Fouling Convention and that non-compliance should result in the detainment or exclusion of the ship concerned.

9
Connections

Now that the international chemical safety system has been outlined, the following chapter will describe the connection between the national and international systems. The institutional structure will first be explained, and then the integration of the aforementioned activities into national legal systems will be detailed. As German authorities actively participate in such activities, and in view of Germany's central position in the chemicals industry, this country will be used as the main example for the interaction between the national and international regulation of chemicals. Reference to the national practice in other countries will be made in specific cases.

National participation

In most cases, the participation of national authorities in global activities on chemical safety has its origin in the country's membership in international organizations (IOs). However, the legal basis for membership may vary. For example, Germany's membership of the OECD is the result of a formal legal act, in which the Federal Parliament (*Bundestag*) enacted a law stipulating the accession of Germany to the OECD.[221] In contrast, Germany joined the WHO in 1951 by means of an act

---

[221] *Gesetz zum Übereinkommen vom 14. Dezember 1960 über die Organisation für Wirtschaftliche Zusammenarbeit und Entwicklung (OECD)* (Act concerning the Convention of 14 December 1960 on the Organisation for Economic Co-operation and Development (OECD)).

of government, without parliamentary participation.[222] Similarly, the participation of German authorities in IPCS activities is regulated by an MoU, concluded by the Ministry of the Environment on behalf of Germany.

There are various types of cooperation between national authorities and international forums, bodies, and organizations. Generally, one authority has the role as a contact between IOs and foreign agencies and the national administration. There are National Focal Points for the IPCS and IFCS, the Competent Authorities for the PIC Convention and National Coordinators for OECD Test Guidelines Programme. In each case, their function is to serve as a contact point for foreign agencies, IOs, or other institutions in the respective country and to synthesize the input of other national agencies. As chemical safety is a cross-sectoral responsibility, in Germany numerous federal ministries and agencies maintain departments that deal with specific aspects of chemical hazards. Coordination among them is necessary to shape and articulate a uniform German position. The Federal Ministry for the Environment, Nature Conservation, and Nuclear Safety plays a prominent role, maintaining a department on international chemical safety (*Referat IG II 3 "Internationale Chemikaliensicherheit, Nanomaterialien, nachhaltige Chemie"*). This department serves as Germany's National Focal Point for the IPCS and IFCS and has been very active in the negotiation of various international agreements.

Participation in these activities requires expert knowledge. Therefore, officials from the more specialized federal agencies attend the specific meetings, discussion groups, etc. For example, experts from the Federal Institute for Occupational Safety and Health, the Federal Institute for Risk Assessment, and the Federal Environmental Agency participate in the OECD SIDS Programme, for which the national contact is the Notification Unit (Feller, Kowalski, and Schlottmann 2005: 61). Officials from the Federal Institute for Risk Assessment were assigned to Final Review Boards in the process of developing CICADs. The head of the German delegation to the UNSCEGHS is an official from the Federal Institute for Occupational Safety and Health; however, delegates from several specialized agencies and ministries, for example, the Federal Ministry of Transport, Construction, and Housing, or

---

[222] *Bekanntmachung der Satzung der Weltgesundheitsorganisation* (Proclamation of the WHO Convention). The publication of WHO Convention is introduced with the remark that the Federal Republic of Germany joined the WHO in 1951.

the Federal Institute for Risk Assessment, and the Federal Institute for Materials Research and Testing, accompany this person. Legally, these officials answer to their superiors – ultimately, the government supervises the agencies and can issues directives. In reality, however, as a result of their participation as experts, these German officials act with a high degree of latitude. The German position is determined in advance by the relevant administrative actors, but with little political intervention. In contrast to this, other delegations are accompanied by professional diplomats, who monitor the observation of the own state's policy.[223]

Legal integration

For EC members, European chemicals legislation permeates national law. Therefore, legal integration happens mostly at the regional, that is, European, level.

Test Guidelines

OECD Test Guidelines are a good example of legal integration at the European level. Regulation (EC) No. 440/2008 lays down the test methods that need to be applied according to Art. 13 (3) Regulation (EC) No. 1907/2006 in order to investigate the intrinsic properties of a substance. According to Recital No. 4 of Regulation (EC) No. 440/2008, the regulation incorporates Annex V of Council Directive 67/548/EEC, which previously stipulated test methods. Annex V of Council Directive 67/548/EEC was almost identical to Annex I C (80) 30 (Spielmann 2004: 142).[224] The Annex to Regulation No. 440/2008 thus frequently states that the respective test method is based on an OECD Test Guideline.

In US chemicals legislation, the test guidelines are laid down in §§796–98 40 CFR (cf. also §790.60 40 CFR).[225] Section 4 (b) (2) (A) of the TSCA defines the endpoints of the test guidelines

Although the Test Guidelines are internationally agreed upon and legally binding in the EC, authorities find ways to deal with inconsistencies or new developments. First, a draft for a new Test Guideline

---

223 Interview 27 April 2004: 42.
224 Interview 15 December 2004: 5.
225 Code of Federal Regulations, Title 40 – Protection of the Environment.

may already have an impact on testing practice, even though it is not yet in force.[226] If amendments to the OECD Council Decision C (81) 30/Final together with a new Test Guideline are likely, national authorities recommend that companies apply the draft guideline when carrying out chemical testing. Here, the aim is to obtain the most precise data possible.[227] Second, some testing methods may be formally equivalent, so different approaches can be used to test a substance in order to obtain data on the same endpoint. For example, the toxicity of a substance might be tested by feeding it to either mice or frogs. A national authority might – perhaps for political reasons or as a result of a trade-off – consent to the adoption of a guideline on frogs, yet prefer the application of a guideline respecting mice because hypothetically it produces more accurate results. At the national level, the government avoids applying the agreed frog guideline by negotiating a deal with the domestic industry to only use the preferred mice guideline.[228]

GHS

Currently, Council Directive 67/548/EEC regulates the classification and labelling of hazardous substances in Europe. But the road in Europe has been effectively paved for an implementation of GHS.[229] The European Commission has proposed a regulation to implement GHS in the EC.[230]

New Zealand is among the first OECD member states to incorporate GHS into its chemicals legislation. In July 2001, the Hazardous Substances and New Organisms Act was amended accordingly.[231] The competent authorities in the United States are currently determining possible ways of implementing GHS, but concrete measures are not yet envisioned.[232]

---

[226] On the comparable concept in regard to the effect of Directives cf. ECJ, Judgement of the Court of 18 December 1997: 7435, paras 40ff.

[227] Interview 15 December 2004, transcript: 24.

[228] Cf. Interview 15 December 2004, transcript: 26.

[229] For an overview of the implementation status of GHS cf. UNECE 2008.

[230] European Commission, COM (2007) 355.

[231] Cf. Hazardous Substances and New Organisms Act 1996, Part 6. Cf. also UNECE 2008.

[232] United States Department of Labor (2008).

Implementation of Conventions: The PIC, POP, and Anti-Fouling-Conventions

Germany has already signed and ratified the PIC and persistent organic pollutants (POP) Conventions.[233] However, they will not be implemented by separate national legislation. Instead, the EC, a signatory to the agreements, has enacted appropriate legal acts to adopt these conventions in order to establish a common legal framework within the EC.[234] The regulations are directly binding on member states, which means that a further adoption is not required (cf. Art. 249 of the EC Treaty). Although the EC did not sign the Anti-Fouling-Convention, as it is not open for signature to regional integration organizations (cf. Art. 17 (1)), a regulation has been created which establishes an EC-wide legal basis for the banning of organotin compounds.[235]

Substance reports

Neither SIDS nor CICADs have direct legal relevance. Generally, the assessment of the risks of a chemical has no immediate legal consequences (with the aforementioned exception of carcinogenic, mutagenic, or reprotoxic substances); however, since substance reports are the result of a consensus of international experts, they are significant and can form the basis for further action. For example, the CICAD on nonylphenol was the only internationally accepted assessment, and served as a template for national assessments.[236] Developing countries and countries with economies in transition rely on the reports as the basis for further measures, including legal action.[237] Consequently, as one expert remarked, "when deciding on a CICAD they both, the developed and the developing countries, should be considered."[238] CICADs, in particular are useful, as they contain a sample risk characterization. Risks also depend greatly on the environmental circumstances, which differ from country to country. For example, controlling vapour pressure or dealing with volatile substances and implementing appropriate safety measures can prove to be much easier in colder climate zones

---

[233] Germany signed the PIC Convention 11 September 1998 and the POP Convention 23 May 2001. Both conventions were ratified 28 August 2000.

[234] Regulation (EC) No. 304/2003 and Regulation (EC) No. 850/2004.

[235] Regulation (EC) No. 782/2003.

[236] Interview 11 May 2004: 6.

[237] Interview 11 May 2004: 7.

[238] Interview (a) 14 October 2005: 13.

than in tropical areas. Nevertheless, a sample risk characterization can give a country an idea of which risks a substance may pose and how these can be countered.[239] Furthermore, if a SIDS concludes that a substance has a "low priority for further work," companies producing the substance may use this as proof that the substance is harmless.[240] If, however, the report concludes that a substance is a candidate for further work and identifies harmful properties, a state or its authorities will have to justify their lack of action in this respect. A country may ignore the outcome of these reports or deviate from the recommendations, but they are required to produce sound reasons for their inaction. Under recent European chemicals law, a "gateway" for substance reports existed in Art. 8 (2) 4th indent of Council Regulation (EC) No. 793/93. According to this provision any "work already carried out in other international fora" could be taken into consideration in the drawing up of priority lists of existing substances. After a new chemicals regime came into effect in Europe with Council Regulation (EC) No. 1907/2006, the OECD explored possible synergies, considering the use of SIDS in the registration process under Regulation (EC) No. 1907/2006 (OECD Secretariat 2008a: Chapter 1, Annex 1.

General practice

Another question is how cooperation between foreign authorities and IOs affects the daily practices of national authorities. Although it is difficult to find empirical evidence demonstrating this relationship, it is likely that it is not only information but also ideas that are being exchanged at international meetings. Such exchanges will not always affect the legal system as a whole, but might influence only administrative conduct. As a result, this type of information exchange is neither regulated nor is it predictable.

Interaction with world trade law

The standards examined in this book are particularly relevant beyond the field of chemical safety. World trade law, especially the Agreement on Technical Barriers of Trade (TBT Agreement), is important in this regard, as it governs the relationship between technical regulations

---

[239] Interview 13 July 2005: 6–7.
[240] Interview (b) 19 July 2005: 19.

and world trade. Under this agreement, states must base their technical regulations and conform their assessment procedures to existing international standards (cf. Arts 2.4 and 5.4). Arts 2.6 and 5.5 of the TBT Agreement also call upon states to participate in the establishment of international standards in order to harmonize technical regulations and conformity assessment procedures. If a state chooses to implement technical regulations that deviate from international standards, it must notify the other members through the WTO Secretariat, pursuant to Art. 2.9 of the TBT Agreement.

Art. 2.2 of the TBT Agreement, which sets out the purpose of the agreement, underscores the central importance of international standards: member states' technical regulations must not constitute "unnecessary obstacles to international trade." This rule is much more restrictive than Art. III (4) of GATT, which prohibits members from implementing regulations that discriminate against foreign goods. While member states are free to regulate trade to the extent that they do not infringe the principle of non-discrimination, the TBT Agreement requires that technical regulations should also be adequate (Tietje 2003a: marginal ns 67–8). In conjunction with the TBT Agreement's preference for international standards as the basis for technical regulations or conformity assessment procedures, this provision also requires that international standards be adequate and do not constitute an unnecessary obstacle to international trade. As a consequence, a state deviating from international standards bears the burden of proving that its own standards meet these two conditions.

The importance of Test Guidelines, GHS, and the UNRTDG would increase if these standards were international standards within the meaning of the TBT Agreement, but although the TBT Agreement requires states to apply international standards, viewing them as important to world trade, it fails to define what actually constitutes an international standard (Schepel 2005: 178; 185). The agreement does address the meaning of "standard," defining it as: "[d]ocument approved by a recognized body, that provides, for common and repeated use, rules, guidelines or characteristics for products or related processes and production methods, with which compliance is not mandatory. It may also include or deal exclusively with terminology, symbols, packaging, marking or labelling requirements as they apply to a product, process or production method..." (Annex I No. 2 of the TBT Agreement). An explanatory note adds that the agreement not only covers standards that are based on consensus, but also includes within its scope those not established by consensus.[241] In addition, the TBT Agreement *only* covers voluntary standards, and requires that the international body or system establishing the standards be open to all member states (Annex I No. 4 of the TBT Agreement).

The matter of international standards remains unresolved. A battle rages between private United States standardization organizations and the European Commission. While the latter argues that the ISO and the International Electrotechnical Commission for the generation of international standards enjoy a monopoly in this area, the American Society for Testing and Materials and the American Society of Mechanical Engineers insist on a more pluralist approach.[242]

In the end, the content of the term "international standard" must be extracted from the TBT Agreement. The relevant provisions in this context are Annex I Nos. 2 and 4 of the TBT Agreement. The language employed in the TBT Agreement does not support the European Commission's narrow position; it does not follow from Annex I No. 4 of the TBT Agreement that only one or two bodies prevail, instead, the provision should be interpreted broadly. In this regard, the TBT Agreement operates under the assumption that several such bodies or systems exist. Moreover, the TBT Agreement requires only that the bodies establishing international standards be open to the relevant bodies of member states of the WTO. Accordingly, any standard issued by such a body can be considered to be an international standard.

The requirement that the international system or body must be open to all WTO members makes it difficult to categorize the Test Guidelines as international standards within the meaning of the TBT Agreement. As has been described above, the Test Guidelines are created by the WNT, a body comprised of representatives from OECD member states, and are eventually to be adopted by the OECD Council, which represents member states. While the OECD has already grown to be more than a regional organization, its membership is nevertheless limited. According to Art. 16 of the OECD Convention, the OECD Council invites countries to join, provided they "prepared to assume the obligations of membership." This is probably not the case for many WTO members. The OECD allows non-members to participate in WNT meetings and concludes treaties with them, integrating them into the MAD system, but this practice does not mean that it is "open" within the

---

[241] Cf. also Report of the Appellate Body, European Communities – trade description of sardines, 26 September 2002, para. 222.

[242] For an overview of the debate cf. Schepel 2005: 186ff.

meaning of the TBT Agreement. "Openness" in this context likely means that states are given the opportunity to participate actively in the decision-making process. This is clearly not the case if states participate merely as observers or choose to implement the Test Guidelines. Thus, despite their relevance and international recognition, the Test Guidelines cannot be regarded as standards within the meaning of the TBT Agreement. Accordingly, any state that requires that chemicals to be tested in accordance with stricter methods than those laid down in the OECD Test Guidelines cannot be reprimanded on the basis of the TBT Agreement.

The situation is different with respect to both the UNRTDG and GHS. These standards are determined by a committee whose composition is decided by ECOSOC. While UNSCEGHS and UNSCETDG are located at UNECE, the membership of these organs is not limited to UNECE members. In principle, every UN member state may participate in UNSCEGHS or UNSCETDG.[243] Since almost every nation state today is already a member of the UN or could easily attain membership (cf. Art. 4 (1) of the UN Charter), one may conclude that UNSCEGHS and UNSCETDG are open bodies or systems within the meaning of Annex I No. 4 of the TBT Agreement. GHS and the UNRTDG regulate the classification of chemical substances, as well as the labelling and packaging of dangerous substances. These are areas that are explicitly mentioned in the definition of standards in Annex I No. 2 of the TBT Agreement. As a result, GHS and the UNRTDG can be considered to be international standards within the meaning of the TBT Agreement.

---

[243] *Supra* 84.

10
Conclusion

As early as 1978, international legal scholars observed a "confusing multiplicity of organizations, each with a narrow perspective on what is essentially a unified threat to human health and the environment" (Alston 1978: 415). Presently, matters are even more complicated, because the number of actors and activities has expanded over the past 25 years. Recent efforts aimed at channelling and coordinating such activities, including the broad strategic initiatives under Agenda 21, the WSSD, IFCS, and SAICM, and the coordinating efforts of the IOMC developed during the 1990s. The IFCS represented an attempt to bring government and private actors together to establish a joint strategy to combat global problems arising from chemical use. Now, although the SAICM may tie in with the IFCS and other approaches, it is conceptually different. One wonders what was so wrong with the IFCS that the need for the SAICM arose. Despite the fact that transnational cooperation is necessary, the chemical governance system seems to have attained an almost hypertrophic status.

It is especially necessary for countries with economies in transition as well as developing countries to take part in the various activities created for their benefit. However, these countries often lack the resources to send officers to each and every committee.[244] As a consequence, the committees and working groups, which were actually created to assist such countries, run the risk of not reaching or even being attended by their target audience. Yet, it is clear that, to quote

---

[244] Interview (b) 14 October 2005: 6–7.

a government official, "the international train is rolling, slowly, but at least it's rolling."[245] There seems to be an awareness among government officials that chemical safety can be attained only through cooperative measures with a special focus on the needs of developing and transitional countries.

---

[245] Interview 11 July 2002: 4.

Part III

The System of Transnational Public Governance

11
Analysing the System of International Chemical Safety

Several specialized bodies make up the structural part of the system of international chemical safety. This chapter will scrutinize these bodies in light of the discussion in Part I on the institutional dimension of global governance. Until now, scholars have identified administrative subunits of the state, civil society actors, and IOs as the relevant players in the international arena. Transnational bureaucracy networks are considered to constitute a new, more effective mode of cooperative interaction in global governance. As a result, the analytical focus will be on the interconnection and the individual role of national agencies, IOs, and civil society actors. The underlying question concerns individual contributions to the resolution of global environmental problems and – most importantly – the actual role of transnational bureaucracy networks. The legal structure of such networks will also be analysed, including an examination of the legal instruments employed and their relationships to each other. The discussion on transnational law, touched upon in Part I, will be picked up again to determine the characteristics of transnational law and its relationship to both national and international legal orders.

Institutional structures

The following section will examine the relevant bodies, including their structure and the participating actors, and their contribution to the overall system of international chemical safety.

IPCS

In Part II, the IPCS was described – along with the IOMC and IFCS – under the general – and tentatively worded – heading of "Interorganizational or Intergovernmental Institutions," since its actual nature is not immediately clear. The IPCS was established through an MoU between the WHO, ILO, and UNEP, and maintains relationships with several countries and their agencies, which are also based on MoUs. The MoU between the WHO, ILO, and UNEP endows the IPCS with a distinctive structure, mainly because it sets up the ICC and PAC as organs involved in the decision-making process and involves government agencies from the countries involved.

The first possible category is the IO's subsidiary organ or subunit. Typically, the so-called principal organs, that is, those organs established by the constituent instrument of an IO, have, by virtue of their constitution or in accordance with the implied powers doctrine, the power to create subsidiary organs to perform their functions. Subsidiary organs can be set up by the principal organs of an IO or even by the principal organs of different IOs. Their roles may be wide-ranging, but they generally provide assistance to the principal organs in the performance of their functions. As they are fully integrated into their IOs' structure and hierarchy and are subordinate to their principal organs, subsidiary organs do not usually have much autonomy (Bernárdez 1988: 108ff.: 140–2; Amerasinghe 2005: 139–41). However, IOs' principal organs may also establish non-subsidiary bodies. Typically, such bodies possess judicial powers and eventually make decisions binding the principal organ.[246]

Several aspects of the IPCS point towards its categorization as a WHO subunit. First, it was created by a WHA resolution. The WHA is a principal organ explicitly mentioned in the WHO Constitution. Second, the IPCS is located at the WHO, with its offices in the WHO building at 20 Avenue Appia, Geneva. The WHO provides secretarial services and a large portion of the program's funds. Its influence is also reflected in its strong role in the decision-making process within the IPCS. The Director-General of the WHO appoints the IPCS Director and the majority of the PAC. Furthermore, the WHO's own PCS acts as the IPCS's Central Unit.

Despite these features of the IPCS, there are indicators that contradict its designation as a WHO subunit. Most importantly, the IPCS is based on a MoU concluded between the WHO, ILO, and UNEP.

---

[246] Amerasinghe 2005: 142; Cf. also ICJ, Advisory Opinion, 13 July 1954: 47 and 61.

The last two participate in the decision-making process through the ICC. The ILO also contributes funds and sends representatives to the PAC. A host of countries have also concluded MoUs with the IPCS (and not just with the WHO) and contribute through the participation of their competent agencies in its various activities. As a result of the tripartite arrangement that forms the basis of the IPCS and the involvement of state agencies, the IPCS is endowed with an autonomy that exceeds the usual leeway of IO subunits. In fact, the IPCS is recognized as an important player in the area of international chemical safety as "the nucleus for international cooperation" (Chapter 19 para. 6, Agenda 21).

This gives rise to the question whether the IPCS is now an IO. International law provides few legal criteria for the classification of an IO. Art. 2 (1) (i)) of the VCLT offers one definition: IOs for the purpose of the convention are any "intergovernmental organizations," that is, organizations created by an agreement between states. However, as stated in the provision, the definition applies to the convention itself, and hence, is only valid in this specific context. Over time, a number of elements have been defined to flesh out the somewhat loose definition laid down in the VCLT. Thus, IOs are commonly defined by the following criteria (Brownlie 2003: 649; Seidl-Hohenveldern and Loibl 2000: marginal n. 105; Köck and Fischer 1997: 60–1):

- They are permanent – although their creation is not irrevocable – associations of states (intergovernmental) based on an agreement under international law.
- They are equipped with organs and empowered to formulate their own will.
- They are bearers of rights and duties under national and international law.

The first criterion, an intergovernmental agreement under international law, excludes institutions created by private law agreements. To such an extent that this criterion specifies Art. 2 (1) (i) of the VCLT, this provision does not immediately make it clear that the organization must have been created under international law to qualify as an IO. Organizations created by private actors may participate in international affairs, but according to this criterion cannot be IOs. International legal scholars have argued that IOs can also be created through "soft law," the decisive factor being the intention of the participating states. This has been exemplified in the case of the CSCE, which started off in the 1970s as a series of conferences. The participating states adopted the Charter of Paris in 1990,[247] which not only reinforced and redefined the objectives of the CSCE, but also endowed it with rights and obligations and provided for institutional arrangements, that is, created appropriate bodies, so that it could meet its mandate (Seidl-Hohenveldern 1995: 231–8). However, neither the Charter of Paris nor the CSCE Final Act (concluded at the first conference in Helsinki in 1975) indicate an intention on the part of the participating states to enter into a legally binding agreement. Nevertheless, their intention was to create an IO (Id.: 230). This intention is also reflected in the Budapest Document of 1994, which renamed the CSCE as the Organization for Security and Co-operation in Europe (OSCE).[248]

Furthermore, an IO should be equipped with at least one organ that is able to formulate and express an independent will (Seidl-Hohenveldern and Loibl 2000: marginal n. 111).

The last, and probably most important, defining characteristic of an IO is the fact that it can bear rights and obligations under both national and international law. States cede parts of their sovereignty to the IO for the duration of their membership. The IO is furnished with these sovereign rights and can exercise them in its own name (Id.: marginal ns 106–7).

This brief summary of the constitutive elements reveals that states are the driving force behind the creation of an IO, endowing it with specific competences to perform its tasks.

The IPCS is not based on an agreement between states. Instead, a MoU between the participating organizations forms its basis. The employed language – particularly, its designation as a "Memorandum of Understanding" – indicates that the agreement is not meant to be legally binding.[249] Thus, neither the agreement between participating organizations, nor those between the IPCS and the states are legally binding. This does not immediately exclude the possibility that the IPCS is an IO, as it has been demonstrated that soft law may also form the basis for an IO, depending on the intention of the parties. In the

---

[247] Charter of Paris for a New Europe, Meeting of the Heads of State or Government of the participating States of the Conference on Security and Co-operation in Europe (CSCE), Paris, 19–21 November 1990.

[248] CSCE Budapest Document 1994, Towards a Genuine Partnership in a New Era, Budapest, 7 December 1994.

[249] For the terminology of agreements cf. Aust 2000: 27–8: 404.

case of the IPCS, the main objective was the improvement of coordination and the pooling of resources regarding activities promoting international chemical safety. The IPCS is explicitly designated as a "programme," stressing the aim of the WHO, ILO, and UNEP to bundle their capacities into a single joint venture. Furthermore, the IPCS is closely tied to the WHO. There is little evidence that the IPCS is supposed to operate as a full-fledged IO. Finally, creating an IO would most likely be an *ultra vires* (beyond powers) act by the WHO. Its Constitution makes no provision for the creation of legally autonomous entities. UNEP, which itself is not an IO and thus does not possess legal personality, cannot endow an organization with the sovereign rights essential for legal persons, because it does not have any sovereign rights to transfer. The definition set out above explicitly requires an intergovernmental agreement. States must provide the legal basis for an IO to operate. Other IOs can certainly become members of an IO, but the fundament must be laid by states. Clearly, this has not been the case with the IPCS. Consequently, the IPCS does not possess the characteristics of an IO.

Despite the fact that it is not an IO, the IPCS has a prominent status. Although it is embedded into the organizational structure of the WHO, the IPCS enjoys a certain degree of autonomy. Its makeup resembles that of actual IOs, after all it maintains, together with the ICC, its own decision-making body. To this extent, the IPCS seems to be more than just a sub-unit of an IO. But from a legal perspective, it is still less than an IO.

Obviously, legal vocabulary does not provide an adequate terminology to describe an institution like the IPCS. This gives cause to consider other categories it could be included among. Possibly, the IPCS' structure may best be described as a transnational bureaucracy network or a node of such a network.

Transnational bureaucracy networks have been defined as "nonhierarchical, informal structures, interlinking national agencies in a specific policy area with the aim of addressing common problems, either by exchanging information, coordinating strategies for action or formulating common rules."[250]

The fact that the IPCS is based on an MoU gives rise to the question of whether an entity that is based on an informal, that is, not legally binding, agreement could itself be hierarchical. Indeed, the MoU functions to organize work and regulate decision-making procedures. In practice, however, affairs are managed not by instructions that are

---

[250] *Supra* 40.

issued top-down, but rather through consensual means. The inclusion of national authorities is characteristic for the IPCS. They participate on the basis of MoUs, but are not legally obliged to contribute. Functions are carried out after prior consultations. No member merely receives instructions from the centre. The MoUs create an organizational structure without establishing a hierarchy. Instead, the IPCS serves as a node, linking several national authorities. Although the exact motivations of the participating agencies may vary, the common overall goal is the maintenance of chemical safety at the global level through cooperative measures. Hence, the structure surrounding the IPCS has the characteristics of a transnational bureaucracy network.

No legal consequences arise when an organization is established as a transnational bureaucracy network. "Network" is a descriptive, not a legal category. Nevertheless, it is remarkable that an organization, which cannot be adequately described in legal terms, still plays a salient role in international affairs.

IOMC

The IOMC also has been described under the heading "interorganizational and intergovernmental institutions." Like the IPCS, it is based on a MoU between several IOs. Other actors, such as national authorities, are involved only upon invitation to the TCGs. The IOMC basically consists of two bodies: a Secretariat, which is located at the WHO, and the IOCC. The roles of the latter are to facilitate communications between the Participating Organizations and promote coordination and cooperation, to avoid the duplication of work and waste of resources.

A possible classification of the IOMC as an IO subunit may be dismissed rather easily. The consensual decision-making process of the IOCC and its shared budget exclude this possibility. Several characteristics inherent to the IOMC also prevent its categorization as an IO. Like the IPCS, it is a body based on an informal agreement between IOs and the IOs that created the IOMC neither possessed the power nor have the intention of creating an IO. The non-commital language of the MoU and the designation of the IOMC as a "Programme" underscore its informal nature.

The IOMC does not meet the criteria of a transnational bureaucracy network. Like such a network, it links IOs operating with a common goal without establishing a hierarchical order. However, national authorities, being the defining feature of transnational bureaucracy networks, play only a minor role in the IOMC and perform no active functions. Instead, they are invited only to consult in IOMC TCGs.

Therefore, the IOMC should be regarded as an institutional arrangement *sui generis* (one of a kind). It is a joint committee or coordinating mechanism for the participating organizations, flowing from the IOs' power to regulate their internal affairs (*Organisationsgewalt*).

IFCS

Questions also arise regarding the nature of the IFCS. The same bodies that established the IPCS, namely the WHO, ILO, and UNEP, were also behind the creation of the IFCS. However, the IFCS is not the product of a MoU concluded by these organizations. Instead, the participants of the ICCS adopted ToR to establish a forum with regular meetings. Based on their use of the terms "forum" instead of "organization" and "terms of reference" instead of "constitution," they did not intend to create an IO. Rather, the IFCS was meant to be "a non-institutional arrangement." It is not an IO.

Neither can the IFCS be categorized as an organizational subunit. The WHO handles secretarial services for the IFCS; beyond this, the IFCS operates autonomously and is not attached to a specific IO.

The question still remains, however, whether the IFCS can be classified as a transnational bureaucracy network. The recurring sessions of the forum suggest that the IFCS is nothing more than a series of conferences. However, it may be constructive to look at whether the IFCS shares the defining characteristics of transnational bureaucracy networks. First, the IFCS has a non-hierarchical organization. This follows from its creators' desire to establish it as a "non-institutional arrangement" and is also evidenced by IFCS practice. The IFCS does not have a top-down structure and it is not able to issue binding resolutions. Second, the IFCS links a host of national agencies operating in the field of chemical safety. Its overall aim is the realization of the goals set up in the six programme areas of Agenda 21, Chapter 19. To attain these goals, the IFCS agreed on certain "Priorities for Action," a to-do list that is regularly reviewed and evaluated. Furthermore, civil society actors and IOs are closely integrated into the IFCS's work.

The IFCS does not merely bring together a large number of actors operating in the field of chemical safety; rather, it creates a forum that allows for close interaction between participants. Thus, it can be classified as a transnational bureaucracy network with the special purpose of formulating and evaluating policies to address the matter of international chemical safety.

CG/HCCS, UNSCEGHS, and UNSCETDG

The next entities to be inquired into are the two bodies connected with GHS. The CG/HCCS was active during the inception and development phase of GHS. The UNSCEGHS is responsible for the further development and the implementation of GHS.

The CG/HCCS was set up by the ILO, WHO, UNEP, OECD, and UNCETDG. Later, a host of other IOs, NGOs, and states joined the work of the CG/HCCS. Its work program and functions were laid down in a set of ToR, drawn up by the IOs participating in the IOMC. The CG/HCCS had a key role in the coordination process. It devised the guiding principles, directed the activities of the focal points and their work groups, and combined the elements submitted by the focal points into a coherent system.

It is clear from the previous analysis of the characteristics of IOs with respect to the IPCS, IFCS, and IOMC that it is easy to eliminate the possibility that the CG/HCCS can also be classified as such. Rather, the CG/HCCS is an organization created by IOs – not by states. It is based on set of ToR – not on a legally binding agreement.

Furthermore, it cannot be regarded as an IO subunit. Apart from the fact that the ILO provides secretarial services to the CG/HCCS, it has no specific attachment to an IO. It operates within the IOMC framework and reports to the IOCC. Despite this reporting obligation, the IOCC has no power to issue instructions on the group's work.

The third possibility is classifying the CG/HCCS as a transnational bureaucracy network. The CG/HCCS has an informal, non-hierarchical structure, based on a set of rules set out in ToR. It offers guidance to the focal points and mediates conflicts, but it does not issue instructions. Therefore, it is more a clearinghouse than a power centre. The dominant group of actors in the CG/HCCS and its working groups was national representatives, heading the CG/HCCS, who led working groups and carried out the majority of activities. Nevertheless, it must be noted that both IOs and NGOs played a vital role in the creation of GHS. The common goal of the participating institutions was the conception of a harmonized classification and labelling system for hazardous chemicals. Therefore, the overall structure established surrounding the CG/HCCS can be classified as a transnational bureaucracy network. In view of its specific role in this process, the CG/HCCS functions as the central node of this network – similar to the IPCS.

The next question concerns the UNSCEGHS (which emerged from the CG/HCCS) and its classification. The UNSCEGHS is located at the UNECE in Geneva, and was created by an ECOSOC resolution. It brings together agency representatives from UN member states. Although IOs and NGOs may participate, they have no voting rights.

The UNSCEGHS is embedded in the ECOSOC institutional structure. It was established by ECOSOC to create commissions, elect ECOSOC members, provide assistance, and determine ECOSOC's rules of operation. In view of this, the UNSCEGHS must be considered as a subunit of ECOSOC.

Earlier it was pointed out that "transnational bureaucracy network" is a descriptive category, as opposed to a legal one. In fact, Slaughter defined networks within IOs as one possible type of transnational bureaucracy networks (Slaughter 2004: 45). Therefore, the legal classification of the UNSCEGHS as a subunit of an IO organ does not preclude the UNSCEGHS from also being categorized as a transnational bureaucracy network.

The UNSCEGHS is a formal body, which is fully integrated into the ECOSOC structure. However, it shares some of the characteristics common to transnational bureaucracy networks. The UNSCEGHS links several national authorities that deal with the classification and labelling of hazardous substances. As a result, the UNSCEGHS and state authorities share the common goal of harmonizing domestic or regional systems to safeguard international chemical safety. Regarding the issue of whether the subcommittee operates within a non-hierarchical structure, the UNSCEGHS is supervised by the UNCETDG/GHS, which is a superior institution. However, the UNCETDG/GHS's function is to coordinate and promote the work of the subcommittee, and, according to the ToR established by ECOSOC, not to get involved in its actual work. As regards the actual process of developing GHS, the UNSCEGHS operates without detailed instructions. It has been already been pointed out that IOs can set up bodies that operate outside a strict organizational hierarchy. These are usually tribunals that operate independently in order to function properly. The UNSCEGHS requires a certain degree of autonomy so that experts can make decisions free from political influence. Particularly in regards to its relative autonomy, the UNSCEGHS can be considered the node of a network of agencies.

The UNSCETDG operates in a similar way to the UNSCEGHS. It was also created by ECOSOC and it is located at the UNECE headquarters. The UNSCETDG brings together experts from state authorities and other institutions with the purpose of developing a harmonized system of regulations pertaining to the transport of dangerous goods. Because of their identical structure and setting, the UNSCETDG can be assessed in the same way as the UNSCEGHS. It is a subunit of ECOSOC. At the same time, UNSCETDG is the node of a transnational bureaucracy network.

Working Group of National Co-Ordinators of the Test Guideline Programme, Task Force on Existing Chemicals and the Joint Meeting

The WNT and Task Force on Existing Chemicals (Task Force) operate within the OECD framework. The Joint Meeting supervises both of them. It endorsed the Guidance Document that governs the test guideline development and revision process and is the final authority on deciding whether a test guideline is to be approved or rejected. Furthermore, it oversees the work of the Task Force on Existing Chemicals and endorses the outcome of SIAM.

Both the WNT and Task Force are integrated into the OECD's organizational structure. They are subordinate to the Joint Meeting, and therefore lack the competence to make autonomous decisions. This means that they are incorporated into OECD's decision-making hierarchy, and thus are in legal terms are subunits of the OECD.

The fact that the Joint Meeting supervises both bodies precludes their classification as a transnational bureaucracy network. The WNT and the Task Force bring together representatives from national agencies that cooperate with a shared purpose. However, as both bodies are subordinate to the Joint Meeting, they have been incorporated into an organizational hierarchy and thus lack the autonomy necessary to be deemed a network. Nevertheless, it is noteworthy that the bodies assemble officials from member states and are not comprised of OECD personnel. These officials primarily answer to their respective state agencies and are integrated into their own domestic hierarchies. The OECD provides secretarial assistance, and rarely intervenes, if ever, in development of Test Guidelines or SIDS.

The prominent role of the Joint Meeting gives cause for consider-ing its status. It has been mentioned that the Joint Meeting is made up of two OECD bodies: the Chemicals Committee and the Working Party on Chemicals, Pesticides, and Biotechnology. While the OECD Council directly created the Chemicals Committee, EPOC, which is a subcommittee of the Council, established the Working Party. Thus, both the Chemicals Committee and Working Party are fully inte-grated into OECD's organizational structure. Legally, they are also mere subunits the OECD. The WNT and the Task Force operate at a working level, while the Joint Meeting deals with strategic issues like coordinating the various programs, and directing and approving their work.[251]

Can the Joint Meeting also be regarded as a transnational bound-ary network or the node of a network? It certainly does not function informally, as is demonstrated by the organizational acts establishing the two bodies that form the Joint Meeting. However, it clearly con-nects with representatives sent by the competent authorities of mem-ber states.

The examination of UNSCEGHS, and the contrasting cases of the WNT and the Task Force demonstrate the crucial issue in determining whether an IO structure is a transnational bureaucracy network. In the end, it depends on whether it is integrated into an organizational hierarchy. It is relevant that, like the WNT and the Task Force, the Joint Meeting is made up of officials delegated as experts by the competent authorities of member states. Like the UNSCEGHS, the Joint Meeting operates as an expert body within the OECD, and thus is not fully integrated into the organization's hierarchy. This is reflected by the decision-making process as regards Test Guidelines or SIDS. The Joint Meeting – by virtue of its status as an expert body – makes final deci-sions regarding the content of Test Guidelines or SIDS. Superior bod-ies like EPOC or the Council merely endorse and implement what the Joint Meeting has previously agreed upon.

As it also interlinks the representatives of various government agen-cies cooperating with a common purpose, the Joint Meeting, like UNCETDG/GHS, can be considered as forming the node of a transna-tional bureaucracy network within the OECD framework.

---

[251] Cf. again Interview (a) 19 July 2005: 12.

Instruments

One central issue of this book is to explore how law can serve as a global governance instrument. While national law can govern only domestic affairs, many have pointed out the weaknesses inherent in international law in terms of its structural design, including complicated and time-consuming lawmaking processes and an inherent inflexibility. Several changes and new developments in international lawmaking aim to enhance the efficacy of international law, so that it a useful instrument for addressing global problems. The same is said of soft law, which bears great resemblance to international legal instruments, but lacks the obligatory character of "hard" law. Transnational public law was introduced as a somewhat nebulous category, the exact characteristics of which have yet to be fleshed out. As a result, one goal of the analysis will be to find out whether any of the instruments discussed in Part II might be classified as transnational public law. In light of this aim, it is necessary to scrutinize the links between these instruments and other legal acts.

Policy formulation

Four elements of the system of international chemical safety might best be described as instruments of "policy formulation:" the 1972 Stockholm Declaration and its supplementary Action Plan, Agenda 21 (endorsed at Rio de Janeiro 20 years later), the Plan of Implementation resulting from the WSSD (created ten years after Agenda 21), and the Bahía Declaration produced by IFCS III in 2000.

If one scrutinizes the language of the Stockholm Declaration, the phrasing indicates a prescriptive content. For example, the principles pertaining to chemical safety read "discharge...must be halted"[252] or "states shall take all possible steps...."[253] Thus, the Declaration establishes rules. In addition, it is the product of a large state conference that calls upon states to act in accordance with the Declaration. This broad state consensus endows it with the necessary authority to be regarded as a type of law. As states are both creators and addressees of the Stockholm Declaration, the first idea might be to categorize it as international law. But this would require the existence of one additional

---

[252] Principle 6 Stockholm Declaration.
[253] Principle 7 Stockholm Declaration.

element: the will of the involved states to enter into a legally binding agreement.

An interpretation of the language used in the agreement may help to identify such a will. In this regard, one notices a deviation from the language usually employed in such treaties. The use of the term "declaration" instead of "agreement" or "convention" already hints that the states lacked the will to be bound legally by the document. This is corroborated by the introductory section of the declaration, in which its central principles are set out: "[the United Nations Conference on the Human Environment] [s]tates the common conviction that...." Obligations certainly do not arise from statements of common convictions. In fact, as an agreement of states lacking the obligatory character necessary for the classification as international law in the sense of Art. 38 (1) of the ICJ Statute, the Stockholm Declaration is a typical example of soft law.

The Stockholm Declaration addresses states, which are to follow the recommendations laid down in the Action Plan. Although the Declaration does not itself create specific obligations, instead formulating principles that serve as the basis for further state measures, it has had an impact on both national and international law. Most notably, it led to the creation of UNEP and the establishment of the IRPTC – the core of UNEP's chemicals programme and the foundation for other international measures promoting chemical safety. As the outcome of a state conference, it is a typical example of soft law.

The next instrument to be classified is Agenda 21. The mode of expression chosen here resembles that of the Stockholm Declaration. The phrase "governments should..." is repeatedly employed throughout the document. Although the word "should" is used instead of the much stronger "shall," which is characteristic of the language of international treaties and indicates a binding obligation, the document has a rather prescriptive character. The intention behind its creation was to formulate desirable results and produce a to-do list covering a variety of areas related to sustainable development. Agenda 21 is one of the outcomes of the UNCED. It has the consent of 172 participating states, adding to its weight and giving it a high degree of authority. On this basis, it can be concluded that this is a law of some kind. Remarkably, Agenda 21 not only addresses governments, but includes within its scope virtually everyone who can contribute to the overall goal of sustainable development: governments, IOs, NGOs, and TNEs.[254] The fact that Agenda 21 does not solely govern relations between states, taking a much broader approach, means that it may be precluded from the category of international law. Agenda 21 employs language that avoids the use of terms that indicate that the parties intend being legally bound. Already the designation of "agenda" distinguishes it from other instruments referred to as "conventions" or "treaties." In addition, the provisions themselves are phrased more as admonishments than binding obligations (i.e. the use of "should" instead of "shall"). Like the Stockholm Declaration, Agenda 21 fulfils the characteristics of soft law.

The impact of Agenda 21 on national and international law was even stronger than that of the Stockholm Declaration. Various other documents, conventions, and other agreements refer to it, referencing Chapter 19 as an impetus for their own creation. For instance, the preamble of the resolution establishing IFCS cites Chapter 19[255] as does the foreword to GHS[256] and the preambles to the POPs and PIC Conventions.[257] Chapter 19 also influenced a number of national measures. Similarly to the Stockholm Declaration, Agenda 21 is the product of a huge international conference, and can therefore be categorized as soft law.

The Plan of Implementation agreed upon by the state representatives at Johannesburg in 2002 ties in with Agenda 21. Again, on the one hand, the language used is prescriptive (for example, "[a]ll countries should promote ..." or "[t]his would include the actions... set out below"), while on the other it lacks the obligatory character generally attributed to international law. The result of a large international conference, it is backed by the necessary authority to be considered law, but does not itself constitute international law. The Plan of Implementation is not addressed to governments alone, but also refers to IOs, NGOs, and TNEs. On this basis the Plan of Implementation must also be categorized as soft law.

One of the outcomes of the third IFCS in 2000 was the Bahía Declaration. Determining its actual character is much more difficult

---

[254] graph 22, Chapter 19, Agenda 21: "International organizations, with the participation of Governments and nongovernmental organizations, should..." or para. 33: "Governments and relevant international organizations with the cooperation of industry should...."

[255] Recital No. 6 of the Preamble of the Resolution on the establishment of an IFCS.

[256] Para. 3 Foreword to GHS.

[257] Recital No. 2 of the Preamble to the POP Convention and Recital No. 7 of the Preamble to the PIC Convention.

than for those cases discussed above. In the introductory part of the Bahía Declaration, its authors, representatives from "governments, international organizations, and nongovernmental organizations from industry, public interest groups, and groups concerned with scientific and labour interests ..."[258] confirm their commitment to the goal of global sound chemicals management. The fact that almost every relevant actor in the field of chemical safety endorsed the Bahía Declaration gives it a high degree of authority.

The solemn language used in the introduction resembles that included in the preambles of other conventions or UNGA resolutions. However, this alone does not furnish it with the necessary prescriptive character. Taking into account the key goals set out in Art. 5, it is indeed prescriptive. Beyond the rather abstract commitment to the goal of global chemical safety, this article contains detailed instructions and timelines for the achievement its goals over the next years or two. It must be added, however, that in comparison with Chapter 19, Agenda 21, or the Stockholm Declaration, the overall tone is much more lenient and the phrasing is more of an admonishment than a command. Therefore, it is difficult to view the Bahía Declaration as law.

States, together with IOs and other civil society actors, authored the Bahía Declaration. It concerns not only interstate affairs, but extends to virtually everyone operating in the area of chemical safety. This is reflected in Art. 2, which addresses many different groups: "we call on governments, industry, public interest nongovernmental organizations, labour unions, scientific organizations, international organizations, and the public to engage and join us in our common efforts...."

There are two obstacles preventing the classification of the Bahía Declaration as international law. First, it was created by a heterogeneous group of actors, and does not exclusively regulate interstate affairs. Second, although the language of the Declaration reflects a commitment on the part of the parties to a common cause, its cautious formulation makes it difficult to infer that the parties intended to be legally bound by it. Its origin also precludes it from being classified as soft law, since soft law arises from commitments made by IOs or states and, as indicated in the introductory text, the Declaration is the product of a cooperative effort of states, IOs, and civil society actors. Because of the presence of these factors, the Declaration does not fulfil the criteria of international law. Instead it should be classified as a policy document

---

[258] Cf. Art. 1 Bahía Declaration.

that reaffirms a commitment on the part of its members to reach certain goals.

The SAICM, which is comprised of the Dubai Declaration on International Chemicals Management, the Overarching Policy Strategy, and the Global Plan of Action, is very similar to the Bahía Declaration. The language of the Dubai Declaration is solemn, authoritative, and official in character. It has been widely endorsed by governments, civil society representatives, and the chemicals industry. But because of its heterogeneous origin, it cannot be classified as international law. The approach taken in para. 28 of the Dubai Declaration, however, is noteworthy: here, the parties "...acknowledge that as a new voluntary initiative in the field of international management of chemicals, the Strategic Approach is not a legally binding instrument." Obviously, the declaring parties found it necessary to make such an explicit statement. In fact, the voluntary nature of the SAICM and, specifically, the Global Plan of Action were discussed in the course of the ICCM.[259] This provision makes a clear statement on the non-legal nature of the SAICM, and reflects the commitment of the declaring parties. Therefore, like the Bahía-Declaration, the SAICM can be classified as a policy document, and not international law.

Standardization: Test Guidelines and GHS

The Test Guidelines regulate the investigation of chemical properties for a specific endpoint. They are developed by the WNT and approved by the Joint Meeting, the latter being the node of a transnational bureaucracy network. Upon the approval of the Joint Meeting, the Test Guidelines are eventually adopted by the Council as amendments to OECD Council Decision C (30) 81/Final.

Prima facie, the Test Guidelines themselves do not constitute law. Rather, they set out the properties of a substance to be investigated, aiming at a particular endpoint, and addressing anyone who carries out such tests. By virtue of the power conferred on them as an OECD body and on the basis of the expertise of their members, the WNT and the Joint Meeting are regarded as authoritative. However, they do not themselves possess the power to require a party to apply the Test Guidelines.

---

[259] Report of the International Conference on Chemicals Management on the work of its first session, paras 26, 42ff.

The Test Guidelines are included in OECD Council Decision C (81) 30/Final. This Decision has two parts. In the first part, the Council has instituted the principle of MAD. According to this concept, any data that has been generated by a member state for a Test Guideline that is in line with GLP principles has to be accepted by the competent authorities of another member state. In the second part, the Council recommends that member states apply the Test Guidelines laid down in Annex I. The Council has already formally adopted the Test Guidelines. In this context, it should be pointed out that, according to Art. 5 of the OECD Convention, Decisions of the Council are binding on all member states, while recommendations of the Council are not. Thus, despite the obligatory nature of decisions, Recommendations are, by nature, less authoritative. Since it is an IO that generates Decision, they have the characteristics of supranational law. Recommendations, on the other hand, fit into the category of soft law. Although the Test Guidelines form a part of Decisions, classifying them as supranational law would go too far. It is a common practice of the Council to amalgamate Decisions and Recommendations, as reflected in the title "Decision-Recommendation."[260] One part of the act is a legally binding Decision, the other is only a Recommendation. Decision C(81)30/Final has two parts. As previously mentioned, the Council in the second part explicitly "recommends" – as opposed to "decides" – that member states apply the Test Guidelines laid down in Annex I.[261] Therefore, the Test Guidelines are soft law. The Decision addresses OECD member states, which have the task of transforming the Test Guidelines into national law, and ensuring their implementation.

The MAD mechanism serves as an incentive to incorporate the Test Guidelines into domestic law; this sets the Test Guidelines apart from classical soft law instruments and those typically associated with IO resolutions. The Test Guidelines are "products" of a technical harmonization process within a transnational bureaucracy network and designed to be implemented by states through legal measures. In contrast to, for example, UNGA resolutions or conference resolutions, the Test Guidelines are phrased in such as way so as to leave little room for

---

[260] OECD Council Decision-Recommendation C (89) 87/Final.

[261] The exact phrasing is: "To implement the Decision set forth in Part I: 1. [the Council] recommends that Member countries, in the testing of chemicals, apply OECD Test Guidelines and OECD Principles of Good Laboratory Practice, set forth respectively in Annexes 1 and 2 which are integral parts of this text."

ambiguity. Their influence arises not only from the fact they have been devised by experts, but also that they are tied to an economic incentive. Because they are sourced in a transnational bureaucracy network, and incorporated into domestic law, the Test Guidelines can best be described as transnational public law.

The UNRTDG and Model Regulations detail provisions on the safe transport of dangerous goods, including chemical substances. The Model Regulations, which form an integral part of the Recommendations, are very authoritative. This is reflected in their designation ("Regulations") and the language employed therein ("shall"). The Recommendations are generated by the UNSCEGHS, approved by the UNCETDG/GHS, and eventually endorsed by ECOSOC.

Like the Test Guidelines, the Recommendations and Model Regulations are prescriptive, but are not issued by a body with the authority to force others to apply them. Instead, an ECOSOC resolution is a type of order to apply these rules. However, ECOSOC resolutions do not bind states: they are soft law and their implementation is not mandatory. However, the longstanding proficiency of the participating experts and the facilitation of trade by uniform rules constitute incentives for the implementation of this system (Jones and Yeater 1992: 310).

The Recommendations and the Model Regulations have had an enormous impact at the domestic level, as demonstrated by their widespread implementation. They resemble the Test Guidelines in many respects (for example, their origin in a transnational bureaucracy network, their non-binding nature, and the strong incentives for implementation that back them). Accordingly, they should also be regarded as transnational public law.

GHS bears great similarity to the UNRTDG and the Model Regulations. The rules on the classification and labelling of hazardous substances originate with the UNSCEGHS, and are later endorsed by ECOSOC. Although they appear authoritative, the UNSCEGHS has no the mandate, nor do they have the power to direct states to apply them. Instead, ECOSOC calls upon states to implement GHS themselves. Again, the status of the UNSCEGHS as a body that convenes government experts, facilitates trade, and promotes the protection of health and environmental safety across the globe constitutes strong incentives for states to follow ECOSOC.

Because of its similarities with the UNRTDG – a transnational bureaucracy network intentionally created it and states are supposed to implement it – GHS may also be categorized as transnational public law.

It is remarkable that all three of the standards presented here have been created by transnational bureaucracy networks, whose nodes were set up by IOs. The development of such standards does not necessarily guarantee their application. In all cases, they are ultimately "activated" through endorsements of IO organs. This is a necessary element that finalizes the procedure and serves the purpose of promulgating the particular measure.

Substance reports

Above, two activities were presented as substance reports. SIDS are basic information sets on the properties of a substance, which are elaborated by an OECD task force in cooperation with the chemicals industry. CICADs are more elaborate and also contain risk information. They are developed within the IPCS network.

Both instruments are collations of substances' properties and exposure data. They represent a consensus of expert consensus opinion on the hazards and risks of particular substances. However, this consensus does not lead immediately to specific regulatory activities.

Substance reports are not prescriptive. Lacking prescriptive content, they cannot be law. It must be pointed out, however, that substance reports may serve as the basis for future action. They may eventually form the basis for legal measures, such as restrictions or bans. Because of their preparatory character, they are legally relevant.

Information exchange

It can be concluded on the basis of its designation, language ("Each Party shall...instance"), and adoption by state representatives that the Rotterdam Convention is a treaty of international law. It emerged from the regimes introduced by the London Guidelines and the FAO Code of Conduct, which established a legally binding PIC mechanism, meaning that these soft law instruments had a preparatory function.

The PIC Convention regulates the exchange of information prior to the exportation of a particular substance. Developing countries and countries with transitional economies, in particular, can make use of this instrument when regulating the importation of hazardous substances. The PIC Convention aims to strengthen the position of such countries in regards to chemical safety. To ensure effective implementation of this agreement in developing countries and countries with transitional economies, it provides for technical assistance advanced by developed countries. Other actors like UNIDO or UNITAR also provide assistance in improving infrastructure and capacity building in these states. Under this regime, implementation is not merely the task of the individual state, but instead the community of states is obliged to cooperate to ensure the agreement is implemented effectively.

The PIC Convention has also responded to past demands for greater flexibility. The Annex, which lists the substances that are subject to the PIC procedure, is not a closed catalogue and the parties can propose amendments to the list. The Chemicals Review Committee, a body consisting of experts designated by the parties, plays a strong role in the amendment process. After reviewing the relevant data, it decides whether or not to recommend an inclusion of a particular substance in Annex III. This procedure ensures that the PIC regime is both flexible and open to new scientific data or socioeconomic development. However, it does not go so far as to allow majority votes in the crucial matter of amending Annex III.

Restrictions and bans

The POP Convention regulates POPs, initially restricting the use of or banning 12 substances. The description of the negotiation process confirms its status as a treaty of international law. With a mandate from UNEP's Governing Council, many states negotiated and concluded this agreement. Its binding character is reflected in its language, designation as a "convention," and authoritative wording ("...each party shall...").

In its design, the treaty responds to the discussion on the inadequacies of international law. First, the POP Convention is a flexible regime. Amendments are possible, even if some parties do not comply or go so far as to obstruct its implementation, since they can be passed without full consent. In fact, with a three-fourths majority, it is possible to override a dissenting party. In such situations, the objecting party may declare that it will not be bound by such an amendment. This opt-out mechanism ensures the continued operation of the POP Convention if controversies regarding measures pertaining to particular substances arise.

Another remarkable element of the POP Convention is the Persistent Organic Pollutants Review Committee, which plays an important role in the amendment procedure. This committee, made up of experts appointed by the parties, contributes toxicological expertise and prevents a bias towards the socioeconomic aspects of restricting or banning a particular substance.

Developing and countries with transitional economies are not ignored in the implementation of the POP Convention. The Treaty requires that developed parties provide financial and technical assistance to states in need. In addition, UNEP and UNIDO have launched programmes to assist less developed countries in the implementation process. Thus, the treaty does not take a stand-alone approach to global chemical safety, but guarantees the support of outside actors and is integrated into a broader strategic scheme.

The POP Convention also addresses the problem that some states might chose to elude the regime by not to becoming party to it. By imposing restrictions on trade with these countries, the convention circumvents the legal doctrine of *pacta tertiis non nocent non prosunt* laid down in Art. 53 of the VCLT.

Two other instruments, the Montreal Protocol and the Anti-Fouling Convention, have also imposed restrictions or bans on certain chemical substances. Both agreements also classify as international law.

The Ozone Convention is considered a classical framework convention, because it only establishes an institutional structure with the capacity to enact future regulations through additional protocols. The Montreal Protocol is one of the most important protocols established within the Ozone Convention framework. It takes into account the special situation of developing countries, allows them to delay their compliance, and provides them with financial support and technical assistance.

It sets up a flexible regime that recognizes the need for further scientific evaluation and provides for cooperative research on the effects of substances on stratospheric ozone. New scientific findings can quickly be incorporated into the existing regime, since a protocol can be amended with only a two-thirds majority. The Montreal Protocol also addresses the problem of free riders by placing restrictions on trade with countries not party to the convention. Prohibiting parties from either importing or exporting regulated substances from countries not party to the Montreal Protocol, effectively reduces the economic advantage gained from free riding.

The Anti-Fouling Convention addresses environmental hazards posed by organotin compounds used to protect ship hulls from barnacles and algae. Because of their harmful effects on the endocrine systems of various marine organisms, several of these substances have been restricted or banned. The Anti-Fouling Convention can be amended by majority vote, and allows objectors to opt out of such new agreements. Thus, the regime is not static, but is flexible enough to adapt to new scientific findings and address new problems. Technical committees are heavily involved in the process of preparing amendments, which adds to the rationality of the decision-making process. Lastly, the Anti-Fouling Convention contains details on non-compliance and places the compliance control into the hands of the parties themselves, who can take effective measures to counter non-adherence.

12
Evaluation

Based on the prior categorization of the empirical data, the following chapter will evaluate the results and contrast the findings with the various aspects of globalization and governance outlined in Part I. The main issues addressed there were how global problems can be regulated and who is responsible for protection of the world risk society. An attempt will be made to answer these questions by looking at the empirical material presented in Part II and the analysis in the preceding sections.

State and administration

The beginning of the Part I emphasized the state's inability to deal with various phenomena of globalization. In the past, the modern state had the purpose of protecting its population, which occupied a certain territory. In the age of globalization, the welfare of a state's citizenry does not depend exclusively on factors within a state's regulatory sphere. More often than not, threats emerge beyond the regulatory control of the state. While the materialization of global risks has given rise to a world risk society, a world government capable of handling such risks is nowhere to be seen. The question stands of whether the emerging structures of global governance can address these risks. The other relevant question in this context is what it might mean for state sovereignty if national authorities form transnational bureaucracy networks.

An attempt to answer these questions can be made on the basis of the empirical findings of this book. First, transnational bureaucracy networks and transnational public law are very common phenomena in the field of global chemical safety, signifying the power of transnational administrative relations. However, it has also become clear that these informal arrangements depend on the establishment of more formal structures to be effective, so that formal and informal structures complement each other.

The use of informal structures indicates reluctance on the part of states to enter into obligatory agreements, in order to avoid further restrictions ofntheir sovereignty. It has been pointed out in the discussion of sovereignty that cooperation among states is actually necessary in order to exercise sovereign rights.[262] Since participation in transnational bureaucracy networks does not entail any legal obligations, states may easily enter into cooperative arrangements without sacrificing their sovereignty. By allowing its administration operate in these transnational settings, the state gains maneuverability. For similar reasons, interstate cooperation leads to the creation of transnational pubic law, as it is informal and not legally binding.

What can be gleaned from the empirical data presented here? A number of conclusions can be drawn from the analysis regarding the state's role in the age of globalization. The existence of numerous transnational bureaucracy networks and the creation of several transnational public law instruments indicate two things. First, the state retains its reluctance to yield its sovereignty. Since no new IO has been created, sovereign rights are not transferred. Legally binding regimes, which by nature restrict sovereignty, are created only if absolutely necessary, for example, in the case of banning or restricting certain substances. When a legally binding regime is established, it is furnished with provisions that conserve sovereignty, for example, by including an opt-out mechanism. Second, the state and its administration both seek ways to cooperate with other states. Regarding the question whether an operation of a national administration at the transnational level erodes sovereignty, this would likely be the case if transnational bureaucracy networks operated autonomously, escaping the control of superior organs and thereby supplanting the function of the primary lawmaker. Although these networks enjoy broad discretion, the domestic hierarchy still exercises considerable influence. Nevertheless, the issue of the legitimacy of decisions made by these networks remains a major concern, which will be addressed in more detail in Part IV.

The incorporation of transnational public law signifies that the state holds these informal instruments in high regard, permitting them to permeate the shell of sovereignty, whereas the metaphor of sovereignty

---

[262] Cf. *supra* 17–22.

as a hard shell must be rejected. Sovereignty is understood more as a membrane, which can be permeated if certain requirements are met.

The state's administration plays a key role in the development of measures in the area of global chemical safety. After all, it is the state's biggest asset for the achievement of certain objectives. In the case of the transnational bureaucracy networks observed in the field global chemical safety, the individual state involves its agencies in solving global problems. The domestic administration itself becomes an actor in global governance, or to be precise, in transnational public governance. This is vividly demonstrated in cases where government officials assume a global perspective, taking responsibility for weaker states, and no longer viewing themselves as mere servants of the delegating state.

In view of these findings, it appears that the state has not really put its sovereignty into the service of global problem solving. While the concept has undergone considerable changes over the past centuries, states still consider carefully whether to confer their sovereign rights on a superior entity. What they are really doing is putting their administration, with all its resources and expertise, at the disposal of the global community.

IOs and transnational bureaucracy networks

The verdict on IOs as contributors to international governance in Part I was rather pessimistic. Some authors regard IOs as being too inflexible to contribute effectively to the resolution of global problems. They were established in the mid-twentieth century by states as a way of cooperating on common issues, at a time when the global pace moved much more slowly. IOs were designed to include decision-making bodies, made up of representatives from member states, an executive body, and a Secretariat. These structures can either be too sluggish to respond to global problems or simply become overburdened, lacking both resources and the ability to expand their competences. Some states view IOs with skepticism or even fear, despite their own membership in such organizations. The main concern is that a powerful IO might encroach upon their sovereignty.

Transnational bureaucracy networks were thought to possess the necessary flexibility to make them a twenty-first-century alternative to IOs. As non-hierarchical entities that bring together members from several national agencies, they are able to operate without a large infrastructure and can access their resources rapidly and easily. The operating costs of such networks are low, as are the sovereignty costs, since states do not commit themselves formally. However, problematic aspects also exist. As a result of their structure, transnational bureaucracy networks lack the visibility of a large IO. Accordingly, their decision-making processes may be secluded and therefore obscure.

First, IOs have found ways of addressing the issue of overlapping areas of responsibility. In the field of chemical safety, they have set up the IOMC as a mechanism for delineating responsibilities and coordinating activities. Furthermore, three of the most important actors operating in this area have established the IPCS to facilitate cooperative activities, creating something that is organizationally connected to the IO – it is administered by the WHO – but functionally separate.

The empirical data demonstrates that IOs and transnational bureaucracy networks are closely related. While the example of IFROs suggested that such networks arise spontaneously and autonomously, the empirical findings demonstrate the strong role of IOs in the emergence of transnational bureaucracy networks. The node of the network is usually embedded in an IO's institutional structure. For example, despite its reliance on services and infrastructure provided by the WHO, the IPCS demonstrates how such a node can operate almost independently.

Actually, IOs set up the node of the network. They provide staff, secretarial services, and an institutional infrastructure. Thus, IOs enable networks to operate. IOs view the establishment of such network nodes within their organizational structure as an extension of their range, as they can source out certain tasks that might otherwise drain their resources or exceed their competences.

The analysis of the standards established in the area of chemical safety has revealed that all of them originate in transnational bureaucracy networks connected to IOs. However, they do not possess sufficient power and accordingly are not endowed with the mandate to call for the applicability of these standards. This task is left to the primary organs of the parent IO.

In view of the empirical data, the question arises of whether the promise of the networks' more effective governance capabilities can be corroborated. It has been pointed out that networks gain their efficacy through flexibility. The transnational bureaucracy networks operating in the area of chemical safety are certainly not rigid structures. However, it is remarkable that their nodes possess the features of an organization. In order to function properly, certain organizational or secretarial duties must be carried out independently. In the case of the IPCS, the node performs these tasks. In other cases, nodes are located within the IO, which carries out these functions. Furthermore, the workflow within transnational bureaucracy networks is structured by MoUs or ToRs, which serve as an internal law framework.

Civil society actors have been integrated into the work of all networks operating in the field of chemical safety. Usually, their involvement does not go beyond the status of an observer, so their attachment to the network node is rather loose. Remarkable in this regard is the role of the chemical industry in OECD HPV Programme, for which it provides the basic information for the substance reports, thereby playing a very active role in this process.

All of the transnational bureaucracy networks examined here are made up of officials from national agencies, who act as experts in the field. As a result, the networks are characterized by immediacy, as many competent persons work together directly to address global problems.

Ultimately, networks only have a minor impact on the sovereignty of the participating states. States do not make themselves subject to a particular regime. Instead, networks have a non-hierarchical structure, and the standards developed within these networks, although backed by an expert consensus, are not legally binding. As a result, no legal obligations ensue from the state participation in networks.

However, networks have also been attributed with a number of flaws that are connected with their rather informal status. One identified weakness of transnational bureaucracy networks is their lack of visibility. Since the networks operating in the field of chemical safety are closely connected to IOs, they are not invisible. Much information – including the creating acts and session protocols – may be obtained via the Internet and at documentation centres of the respective IOs.

The non-obligatory status of networks has been highlighted as an advantage of transnational bureaucracy networks, because states can cooperate together in this setting without sacrificing much of their sovereignty. However, this feature of networks is also a flaw. While IOs can always fall back on their own resources, as limited as they may be, the success of transnational bureaucracy networks rests entirely on the assets contributed by the participants. If these prove to be insufficient, the network has to cut back its activities. For example, OECD HPV Programme has been cut back to the effect that it is now just a collection of data relevant to hazard assessments. Initially, its scope was broader. The HPV Programme began more ambitiously as a risk assessment initiative, but members were not willing to provide the necessary resources.

Civil society actors

The global importance and influence of civil society actors is undisputed, but is not reflected in international law. In transnational public governance, civil society actors play a very limited role. NGOs and TNEs are usually involved as observers. Only in two cases have TNEs played a significant part. The first is the participation of the chemicals industry in the OECD's HPV Programme, where it collated data for SIDS. This is, in fact, the only case encountered in the area of chemical safety, where a private governance measure, the ICCA HPV Challenge, is coupled with public governance. Experts delegated from NGOs and TNEs also participate in SIAMs. Apart from that, private governance measures are carried out independently from public governance.

The second case where TNEs are involved beyond the role of observer is the IPCS' CICADs Programme. In the process of developing a CICAD for a particular substance, manufacturers of a substance are given the opportunity to voice their views on its hazards and risks. In view of the preparatory function of substance reports for regulatory actions, they are granted a hearing.

Civil society actors thus participate in public governance activities. Measures of transnational public governance benefit from the involvement of NGOs and TNEs, but do not appear to depend on them. Nothing in the field of chemical safety indicates that civil society actors could substitute for public governance measures.

Law

Analysis of the empirical data has revealed that many of the instruments employed in the area of chemical safety are indeed law, but, apart from a small number of "hard" law conventions, there are several soft law instruments in existence that fit the definition of transnational public law. The question now is of how these relate to the role of law in the age of globalization.

International law

The fact that only a few of the many instruments employed in the field of chemical safety are "hard" international law is remarkable. Several other aspects are however also notable. First, international law is employed only for a certain type of measures, such as information exchange and restrictions or bans. Information on issues of chemical safety can be disseminated through a variety of channels. The Rotterdam Convention formalizes the exchange of information, replacing two soft law instruments. The latter – the London Guidelines and the FAO Code of Conduct – had a number of flaws, including that they failed to establish infrastructure and support mechanisms for developing countries. As soft law instruments, their enforcement rested on soft power. Consequently, the compliance rate with this convention was rather low. The regime established by the PIC Convention is not without flaws either, but its regulatory architecture attempts to remedy the shortcomings of the preceding instruments.

Three conventions – the POP Convention, the Ozone Convention, and the Anti-Fouling-Convention – restrict or ban certain substances. The small number of restricting regimes and their high degree of specialization – each convention is limited to very specific types of substances – is the result of two factors. First, restrictions and bans are the most incisive tool for maintaining chemical safety. The practice in the EC demonstrates that these measures are used cautiously.[263] The use of a substance is only restricted if its risks have been clearly established and after certain socioeconomic factors have been considered. Restrictions or bans are only imposed if all relevant data are made available, the risks clearly defined, and the conflicting interests reconciled. As the problem of "Toxic Ignorance" remains unresolved, gaps in data prevent a thorough risk assessment. At a global level, conflicting interests do not merely exist within a single state or economic area, but arise across states, such as, for example, between North and South or developed and least-developed states. DDT is a good example of how a substance can pose severe risks and at the same time have enormous benefits. DDT accumulates in the polar environment, but it is an effective pesticide to combat disease vectors in the equatorial zone. Drawing up an instrument of international law that restricts and bans certain substances is a

---

[263] Council Directive 76/769/EEC restricted 47 substances or substance groups.

complicated process that is usually initiated only when pressure reaches a certain threshold. This explains why there are only a few highly specialized conventions regulating this area.

Second, although soft law measures or voluntary action on the part of the chemical industry to abandon the manufacture of certain substances might be considered alternatives to "hard" law solutions, they have many serious disadvantages.

Although states do comply with soft law, such compliance must be achieved through the use of soft power; participants in soft law regimes cannot resort to the enforcement mechanisms that are available in international law. However, in such cases compliance is even more important to the regime's success, because non-compliance might pay off economically. It might prove to be difficult to exert enough soft power through persuasion or economic incentives to achieve adherence to the regime. In this case, free riders might exploit the situation, simply ignoring the regime for their own economic benefit. Voluntary measures taken by the chemicals industry also face the problem of non-compliance. They also must deal with the problem of free riders, who may take advantage of the situation by continuing production, trade, and use of the restricted or banned substance, thereby thwarting the success of such a regime.

The conventions regulating chemical safety make use of the elements introduced in recent years to enhance the efficacy and flexibility of international law. As framework conventions they are furnished with amendment procedures allowing for majority votes and opt-out mechanisms. This guarantees the operability of the regimes, even if parties chose to reject an amendment. Furthermore, the conventions take into account the special circumstances of developing countries and countries with transitional economies. Under these regimes, countries receive financial support and assistance in technical and administrative matters. Additionally, in cases of substance restrictions or bans, time limits are extended for these countries to achieve compliance.

A striking feature of the conventions operating in this area is the expert bodies that they establish. The discussion of international law in Part I pointed out that one way to enhance the efficacy of international conventions is to allow the treaty secretariat the authority to interpret treaty provisions. The idea behind this feature is to depoliticize conflicts regarding the treaty by referring them to an impartial body. Expert bodies, which prepare the decision-making in the amendment procedure, serve a similar purpose. These authorities are not entirely free of political influences. State parties to the convention dispatch these experts, so it cannot be ruled out that they are bound by instructions handed down by their respective government and function more like civil servants than as experts. Nonetheless, these bodies lend both expertise and objectivity to the amendment process, which can often be politically loaded due to conflicting socioeconomic, health, or environmental interests.

Although the scope of the role varies between these agreements, another remarkable aspect of these conventions is the role they assign to civil society actors. For example, the Rotterdam Convention recognizes their importance in this field only insofar as they are admitted as observers, whereas they play a more active part in the POP Convention and the Anti-Fouling Convention. The latter two agreements recognize the unique potential of NGOs for contributing to the efficacy of the respective regimes. By involving them in the amendment process, they are able to draw on the particular expertise of the NGOs. However, the latter are not granted a position equal to that held by states.

In sum, one can conclude that international law is only rarely relied upon to maintain chemical safety, and where no other promising mechanisms are available. Thus, international law has its place in global governance, provided that actors are aware of its shortcomings and benefits and furnish the regimes accordingly. Some scholars were too rash in predicting the demise of international law in the age of globalization. The state and its administration are still an indispensable element of global governance, as is international law.

Soft law

In addition to international law, international actors also make frequent use of soft law. The type of soft law used here consists of agendas, issued by large interstate conferences. Agendas, such as Agenda 21, function as to-do lists, setting out the goals to be achieved and outlining the significant waypoints. They provide the basis for further action, integrating different measures into a coherent package and promoting chemical safety at the global level. These are constantly being evaluated and, if necessary, reformulated.

Transnational public law

The analysis above revealed that a number of instruments applied to promote chemical safety on the global level fall within the category of transnational public law. These are instruments that have their origins in transnational bureaucracy networks and are endorsed by an IO. In the field of chemical safety, transnational public law has been employed to harmonize standards and regulate important aspects of the risk assessment and risk management process. Harmonization guarantees an equal and high level of protection, while promoting economic aims by facilitating the exchange of data and goods across borders.

One must distinguish form and content when examining standards. Transnational bureaucracy networks elaborate the content – the actual standards. They have the necessary technical competence and authority to create such norms. IOs implement the standards through resolutions, which serve as recommendations or endorsements, or, at the very least, serve to publish the standards. According to their form, the standards are soft law, like any other resolution issued by an IO. But they are unique in the sense that they combine the efforts of transnational bureaucracy networks and IO resolutions.

Transnational public law is not self-executing. This is especially true of standards, which interact with national and international laws. The national and international legal order absorbs these standards, thereby declaring them applicable for actors within the respective regulatory area. Transnational standards are normally incorporated into national law or the equally important laws of regional integration organizations like the EC. However, they can also subsumed into international law. The UNRTDG, for instance, have found their way into many conventions pertaining to international transportation.[264] When transnational public law is incorporated into international treaties, often a second transposition becomes necessary. Because international law does not address civil society actors, states agree to implement these standards into the body of national law by signing international treaties that incorporate these norms. This transposition does not always occur verbatim. As with GHS, most standards allow for some leeway in their transposition. However, this latitude is rather narrow and must be carefully exploited to achieve the aim global harmonization.

Transnational public law can also be described as protolaw. It consists of sets of rules that reveal their effect not upon their creation, but upon their incorporation into a legal order. In conjunction with the respective legal order, the standards become applicable and govern the activities of the manufacturer or trader of chemical substances who test, classify, label and package the chemicals they manufacture or transport – in accordance with the standards. The legal systems that are

---

[264] *Supra* 100.

most relevant to them are those domestic legal systems of the states in which they operate.

In view of the informality of transnational public law, it is almost surprising that civil society actors do not play a greater role in comparison with their function international lawmaking. NGOs are often admitted as observers; they represent those parties affected by the standards (or are themselves affected) and often also play a role in the implementation of the new standards. However, neither the IFCS nor the nodes of the transnational bureaucracy networks allot voting rights to NGOs or TNEs. Although their importance is recognized, they have not been placed on an equal footing with other international players.

Recognizing how transnational public law is created and interacts with other legal systems helps to resolve questions of legal theory: does the existence of transnational public law signify the disintegration of other legal orders or does it support the notion of interlegality, that is, the enmeshment of legal orders resulting from their porosity?

The results of the empirical study support the use of a wide definition of law. If approached with the mindset of a legal positivist, the impact of many informal legal activities could not have been uncovered, and their implications for national law unexamined. Teubner posited that global law would emerge from the societal rims (Teubner 1997: 7; 13).[265] This notion should be amended to include the idea that global law, that is, rules that govern global contexts, also emerges from transnational bureaucracy networks. Civil society actors may influence the creation of transnational public law, but ultimately the creation of law is a public effort. Noteworthy is the absence of mixed systems of norm creation. The examples of GHS and OECD Test Guidelines show how public actors allow private ones to participate in the process of norm creation. However, within this process, private actors may only voice their opinions or concerns: state representatives make all final decisions. This signals that the role of the state is not actually diminishing. The areas regulated by transnational public law touch upon those belonging to the core tasks of the state, including the protection of health and the environment.

The different instruments regulating chemical safety that fall within the category of transnational public law do not appear to be fragmented or incoherent. In fact, the agendas in this area serve as a kind of super-structure, which frame the relevant problem areas and tie the various

---

[265] Cf. also *supra* 57.

measures together. A competition or collision of standards could not be detected within this particular field and the empirical investigation also did not reveal a competition or conflict across fields, for example, between standards of chemical and medical safety. Of course, this does not mean that such collisions do not occur. Such conflicts, however, do not reach a level that compromises the system's integrity.

It would appear that transnational public law does not lead to conflicts across different systems of norms. Instead, it shows how different systems mix. This finding supports Sousa Santos' concept of interlegality.[266]

Legal systems are permeable. This is particularly true of national law. Each system performs a specific function, which is derived from its particular characteristics. Transnational bureaucracy networks develop transnational public law specifically with the view that it will have to be integrated into a domestic legal order to become effective. This may lead one to conclude that domestic legal orders take a higher position; however, this initial assessment is not fully supported by the facts. Backed up by the state's monopoly on physical force, the domestic legal system is the suitable forum for immediate influence on the behaviour of civil society actors. Nevertheless, to achieve a maximum effect with regard to the resolution of global problems, individual national measures must be consistent with each other. The harmonization of standards at the transnational level through transnational bureaucracy networks plays a key role in this regard. Measures are synchronized at the transnational level, their implementation is made obligatory for states at the international level, and, if necessary, is ultimately executed at the national level.

---

[266] Cf. *supra* 57f.

13
Conclusion

The empirical data shows that it is both incorrect and premature to declare the state incapable of dealing with global problems. The state's power is not diminishing. Instead, the state reconstitutes its power and seeks other ways to apply it. Weiss describes the state as "catalytic" as it no longer exclusively relies on its own resources, but instead builds alliances with other states and non-state actors to achieve its goals (Weiss 1998: 209). Others describe the state's new role as that of a node of a complementary world system (di Fabio 2003: 76) or as an "interdependence or interface-manager," which bundles and organizes governance resources (Herberg 2005: 33; Messner 2002: 28).

Remaining reluctant to yield sovereign rights and create new formal institutions, the state instead allows its administration to operate transnationally – in informal structures requiring no legal commitment. This predilection for informality, however, does not mean that the era of IOs as formal institutions has ended. The global chemical safety system consists of closely interacting formal and informal structures. The informal structures depend on the existence of formal structures. IOs provide a central operational node for transnational bureaucracy networks to function. Thus, transnational bureaucracy networks are not spontaneous, free-floating arrangements. Instead, they owe their existence to IOs, which initiate, facilitate, and moderate their creation and implementation. As a result, there is no reason to assume that informal structures like transnational bureaucracy networks will displace formal ones such as IOs. Instead, they appear to supplement each other in their respective functions.

The members of transnational bureaucracy networks wear two different hats. From a strict legal point of view, they are civil servants who have been dispatched to an IO subunit. They are bound by the instructions handed down by their superiors and subject to the laws governing the civil service. A holistic legal-sociological approach reveals that these civil servants operate in networks that are attached to the formal structure of the IO but work almost autonomously. Here, the focus lies on the civil servant's expert knowledge, which they contributes to the network.

The close enmeshment of formal and informal structures is also revealed through an examination of the activities of both IOs and transnational bureaucracy networks. The Test Guidelines are one example of how transnational bureaucracy networks can contribute to the development of technical standards. The dissemination and implementation of these regulations is still dependent on formal structures: the decision of the WNT is picked up by OECD bodies and eventually adopted by the OECD Council.

This enmeshment also extends to matters of law. From a legal positivist's perspective, technical standards elaborated by transnational bureaucracy networks can be dismissed as soft law. But the instruments developed by transnational bureaucracy networks are much more. They may be regarded as transnational public law, representing a consensus of expert opinion that helps to predefine further national or international legal provisions, which will eventually be incorporated into international law or national legal systems.

The devolution of lawmaking processes to the transnational level has the advantage that experts can develop provisions in an environment that is less influenced by political constraints. However, this approach can also be problematic if decisions at the transnational level supplant the decisions of national lawmakers, leading to issues of the legitimacy of transnational public law.[267]

Transnational public law serves its own purposes. It is suited to highly technical matters, whose regulation relies on the expertise of government officials. As a result of their technical nature, such regulations are subject to change at a moments notice. Transnational public law, with its basis in transnational bureaucracy networks, provides the required flexibility for such revisions.

---

[267] This aspect will be discussed in detail in *infra* Part IV.

The obligatory character of international law means that it retains its purpose, despite the need for increased flexibility in the age of globalization. In the area of chemical safety, international treaties are used mainly to restrict or ban substances. The regulation of chemicals cannot be based on informal measures alone. Nevertheless, the restriction and banning of such substances also has a highly technical component, and a timely response to new scientific findings is often necessary. The regimes explored here take this into account, incorporating features within the system to enhance their flexibility and effectiveness. Designed as framework conventions, they are open to revisions in principle. Expert bodies are set up, which play an important role in the amendment procedure. The advantage of such bodies is that the amendment process becomes less politicized, and is thus accelerated. These regimes also provide technical and financial assistance to developing countries and countries with transitional economies to ensure the implementation of the treaty and overall performance.

Each legal layer serves its specific purpose. International law is, by virtue of its obligatory character, highly visible. Because of the obligations imposed on states and the possible legal and political implications of every breach, states are prompted into action with these obligations. Transnational public law is both informal and flexible. National civil servants with expertise in these areas (as opposed to plenipotentiaries) contribute to the development of this type of law. The benefits of this process are that these experts can rapidly adapt to changes arising in this field and domestic law is backed by the full power of the state.

Table 13.1 shows the completion of the preliminary typology of legal orders as described in Table 5.1. Supranational law remains a rare exception, the only functioning example being the EC. Allowing an IO to issue regulations legally binding not only on states but also their citizens comes with high sovereignty costs. Sovereignty costs for international law are much lower, but this also means that it has less ability to influence. Soft law and transnational public law both come with low sovereignty costs, but the investigation of the instruments employed to tackle the problems of global chemical safety show that transnational public law has a much higher significance than soft law. It is created already with the view to its incorporation into national legal orders. The major benefit of national legal orders is that the state can effectively enforce them and that creating laws does not incur any sovereignty costs. The

Table 13.1   Typology of law in the age of globalization

Features type	Creators	Addressees	Interlinkage	Sovereignty costs	Examples
Supranational law	IOs (with States)	States; Societal actors	National law	High	REACH regulation
International law	States	States (occasionally, IOs)	National law	Middle-low	Stockholm Convention
Soft law	States; IOs	IOs; States; Societal actors	International law; Transnational public law; National law	Low-none	Agenda 21; Bahía Declaration
Transnational private law	Societal actors	Societal actors	International law; National law	None	Lex Mercatoria
Transnational public law	Transnational bureaucracy networks; IOs	States; IOs	International law; National law	Low-none	GHS; Test Guidelines
National law	State institutions	State institutions; Societal actors	International law; Transnational law	None	TSCA

downside of national law – its limitation to the state's territory – is mitigated through harmonization at a transnational level. By absorbing transnational public law, national legal orders become an element in the concerted action of states in combating global problems.

Part IV

The Legitimacy of Transnational Public Governance

14
Concepts of Legitimacy

Before looking at the legal concept of legitimacy, the following section will provide a short overview of the prevaling theories operating in the areas of sociology and political science.

Legitimacy as a sociological and political concept

Basically, any exercise of authority requires a justification. This proposition follows from the idea that every human is free. But as *Jean-Jacques Rousseau* observed, "man is born free, but everywhere is in chains" (Rousseau 2004: 5). The question arises as to whether and how the shackling of man can be justified, that is, more abstractly, why must a person submit to authority at all (Bodansky 1999: 601).

In order to understand better the close relationship between legitimacy with authority, the meaning of "authority" must first be examined. Authority implies the claim of its bearer that his or her decisions are binding. Thus, legitimacy and authority are intertwining concepts. Since the Middle Ages, legitimacy has been an inherent feature of authority, distinguishing it from tyranny and usurpation (Quaritsch 1987: marginal n. 1990). Others have to subject their will or judgement to the bearer of authority (Green 1998: 584). This close connection to legitimacy distinguishes authority from mere power: People will accept state authority because they perceive it as legitimate, whereas they will comply with an illegitimate tyranny because they fear oppression and abuse (Weber 1978a: 53–4; Würtenberger 1986: 344, Fn. 4).

The following section will provide a short overview of the concept of legitimacy from a perspective of sociological and political science. These concepts cannot be applied directly to investigate the legitimacy of transnational public governance, but they can point to the problematic aspects of legitimacy in regard to globalization, serving as reference points that help to identify problematic aspects in the legitimacy of law.

Sociology aims to describe the reality of human and societal interaction (Weber 1978a: 4). Thus, sociological research focuses on the motives that compel humans to accept governmental authority (Zippelius 2003: 124). In this way, sociology has an empirical understanding of legitimacy and approaches it as a factual phenomenon (Luhmann 1997: 27; cf. also Röhl 1987: 176–7).

Max Weber's three types of authority describe the motivations that compel people to accept authority, and this theory remains among the most influential to date. Weber has observed three ideal types of authority and distinguished corresponding types of sources of legitimacy for each:

- *Traditional authority* is authority, taking the form of government or a set of rules, which is legitimized because it has always existed that way, for example, a monarchy.
- *Charismatic authority* arises in cases where a government's authority is accepted because of its charisma, sometimes on the people's belief that the authority figure possesses special powers, for example, a tribal chief.
- *Rational-legal authority* relates to instances where a government's authority is justified because it is derived from, and is in conformity with, a previously agreed upon or imposed set of rules and procedures, for example, a constitution (Weber 1978a: 215ff.).

From today's perspective, Weber's concept might be considered as outdated, because it does not sufficiently take into account the role of democracy as a source of legitimacy. In modern societies, the single most important source of legitimacy is democracy.[268] Under Weber's approach, democracy plays only a minor role. His approach is empirical, not normative. Yet, it is remarkable that he mentions democracy in conjunction with charismatic authority. The charismatic leader seeks the approval of the masses, thereby reinforcing his legitimate claim to authority (Weber 1978a: 266). Democracy is not considered as an element of rational-legal authority, as one might assume, but is regarded only as a by-product of charismatic legitimacy (Schliesky 2004: 155). However, a modern state, founded on a pluralist society and furnished

---

[268] Beetham 1998; Scharpf 1993: 165; Bodansky 1999: 599; Kaiser 1971: 706.

with a democratic constitution, cannot be based on charismatic authority (Sternberger 1967: 120).

Luhmann has defined another influential sociological approach to legitimacy. In his opinion, the political system of a complex industrialized society can no longer base its legitimacy on universal legal convictions or beliefs. Instead, it derives its stability from different approaches that are constantly in flux. It is necessary, therefore, to identify the structures and procedures that translate the system's variability into stability. A complex system can be legitimized only through procedures, as these produce and reduce alternatives and thereby guarantee the system's acceptance (Luhmann 1997: 251–2). In Luhmann's view, procedures are not ritualized successive courses of action, but rather social systems with the purpose of preparing a binding decision (Id.: 38ff.). Typical cases of such procedures include court proceedings, political elections, lawmaking, and administrative procedures (Id.: 55ff.; 137ff.; 201ff.). Universally acknowledged rules provide the framework for the procedure. These rules assign certain roles to the parties involved. By subscribing to the procedural rules, the parties also agree to accept any outcome as binding – despite the fact it might not favour everyone involved (Id.: 102ff.).

A clear weakness of Luhmann's observations is his disregard for the motivations, ideologies, or worldviews of those subject to the authority, stripping the concept of legitimacy of all references to values or beliefs. Anything can be legitimate, as long as it moves through the proper channels. Values, however, play a significant role in the acceptance of authority (Schliesky 2004: 158; Kopp and Müller: 1980: 107). They are, after all, the reference system of authority. Without them, authority becomes an end to itself (Schliesky 2004: 159). Despite the flaws of his approach, Luhmann brings to mind the difficulties of ensuring legitimacy in complex modern societies.

In the 1960s, Schelsky pointed out how rapid technological and scientific developments impact society, and in the end, and very rapidly, change it. He observed how science and technology enable humankind to loosen the shackles of nature. But as people increasingly use science and technology to shape their world, they become subject to the restrictions science and technology (Schelsky 1961: 17–18). This has consequences for the way societies are governed. Authority and legitimacy are no longer founded on democracy, but on the inherent necessities of science and technology. Political values are replaced by scientific necessity. Modern technology requires no further legitimization, despite the fact it is used to govern. Hence, according to Schelsky, the modes of governance are dictated by science and technology. The will of the people becomes obsolete (Id.: 21–9).

After forty years of advances and setbacks in science and technology, Schelsky's observations still carry some weight. However, it cannot be unequivocally stated that technological or scientific necessity has replaced political values. For example, embryonic stem cell research and therapeutic cloning might be perceived as absolutely necessary from a medical perspective. Nevertheless, these techniques have sparked intense debate on bioethics in many countries. Genetic engineering and advancements in nuclear energy are met with skepticism and face heavy opposition, even though they promise progress and possible solutions to societal problems. The existence of such debate shows that people still hold ethical values in high regard, viewing new technologies skeptically, even though they might be more effective or efficient. Nevertheless, Schelsky's approach points to expertise as an element of legitimacy and thus might be helpful in investigating the legitimacy of transnational public law standards.

Political science distinguishes two types of legitimization in a democratic state. Scharpf termed these input legitimacy and output legitimacy. The former refers to *government by the people*, while the latter concerns *government for the people* (Scharpf 1999: 16). Both concepts support, supplement, and amplify one another in the modern state (Id.: 21).

In the case of input legitimacy, authority is legitimized because it is founded on the will of the people. Input legitimacy usually rests on the participation and consensus of the people. This can, however, lead to problems. The argument that participation falters if the gap between those affected by authority (the governed) and those exercising authority (the government) becomes too great. Furthermore, consensus cannot always be achieved, especially when political solutions do not benefit everyone. Majority rule is a pragmatic solution that requires further justification. Why should one yield to the will of the majority? The theory of input legitimacy assumes that a collective identity – based on a common ethnicity, language, culture, tradition, or set of values – enables the individual citizen to accept majority rule (Scharpf 1999: 17–18). However, input legitimacy becomes problematic in supranational contexts. The EU, for example, has no homogeneous citizenship. Even though politicians stress the common European heritage, a collective identity similar to that within a national state has not yet emerged (Id.: 19).[269]

---

[269] But cf. the conceptualization of a European polity made up of multiple political demoi by Weiler 1997: 118–122.

The concept of output legitimacy focuses on the outcome of political decisions – "government for the people." In contrast to the concept of input legitimacy, which merely seeks to transport and transform the will of the people through political processes, output legitimacy takes into consideration the quality of the political achievements – their furthering of the common welfare (Scharf 1972: 21; 25). *Government for the people* derives its legitimacy from its ability to solve problems that cannot be addressed by other means. Output legitimacy requires stable and multifunctional structures, and also requires a defined political entity, which does not necessarily have to be as close-knit as the demos of input legitimacy (Scharpf 1999: 20–1).

Output legitimacy can be based on a variety of mechanisms. Elections are the first such mechanism. In contrast to their role in the context of input legitimacy, elections here do not serve the purpose of transferring the will of the majority, but instead create a political infrastructure for political accountability. In order to fulfil this purpose, they need to be supported by a system of checks and balances, as well as open discourse (Id.: 23). Another important mechanism is independent expertocracy. Certain types of political decisions are withdrawn from politically responsible officials, and are given over to independent experts. This is especially the case with technical decisions. Instead of being accountable to the public, experts answer to the scientific community, who criticize the decision-making process. If the resulting outcome does not satisfy the electorate, it can always override expert decisions (Id.: 23–4). Furthermore, open, pluralist policy networks, whose activities precede or follow parliamentary decisions or intergovernmental negotiations, can contribute to output legitimacy (Id.: 26ff.).

Although, this overview of legitimacy in the social sciences is not exhaustive, it does reveal several problematic aspects of legitimacy in the current context. First, legitimacy is difficult to maintain in the age of technological feasibility. For Luhmann, technology complicates society so that universal concepts of legitimacy can no longer exist; instead, procedures convey legitimacy. Schelsky has emphasized that technological necessity supersedes the concept of democratic legitimacy. However, neither concept provides a full account of the concept of legitimization, especially since they broadly neglect the role of values in ensuring legitimacy. Second, legitimacy is difficult to maintain in an age of multilayered governance. Scharpf argues that input legitimacy cannot effectively function in a supranational environment. Instead, it must be supplemented by output legitimacy, where the practicability of decisions with regard to the common welfare becomes the decisive factor. From a legal perspective, the operability of these legitimacy concepts might be considered questionable. However, they will prove to be useful as a background to the further investigation of legitimacy.

Legitimacy as a legal concept

In legal studies, particularly in constitutional law, legitimacy is a key characteristic of government authority. The legal approach to legitimacy, as opposed to the sociological understanding, is not concerned with the factual or empirical acceptance of authority, but primarily addresses the acceptability of authority according to legal principles. Of course, both approaches are not strictly distinguishable, being closely intertwined. After all, legitimacy is an interdisciplinary concept that must take into account the research in other areas (Schliesky 2004: 160).

In the modern state, legitimacy as a legal concept is closely linked to that of democracy. According to modern understanding, sovereignty rests with the people. As a consequence, the bearer of authority and those subject to the authority are one and the same (Id.: 163; Kelsen 1966: 326). Therefore, any political authority must be derived from the people and government – in the broadest possible sense – must have the general acceptance of the governed (Böckenförde 2004: marginal n. 3).

The demos confers authority on the government, and, at the same time, legitimizes it. This has the effect that authority can – and must – always be traced back to the sovereign, that is, the people. The demos – usually a group of persons sharing a common citizenship – transfers authority via a specified decision-making procedure.

Since a pure form of direct democracy is impractical, most democratic systems employ an indirect democratic form. Citizens elect representatives and transfer authority to political organs,[270] which must always remain connected to the people. Thus, the state is organized from the bottom up, so state organs are not detached from its citizens. Furthermore, state organs can maintain their representative character only if they preserve their independence. Authority is not transferred permanently to state organs, which are accountable for their actions to the people (for example, representatives in parliament), or other organs representing them (for example, the government answers to the parliament). Regular elections ensure a review of the exercise of authority and represent a constant renewal of legitimacy (Böckenförde 2005: marginal ns 16–9).

---

[270] Cf. Böckenförde 2005: marginal ns 1–12 for a fundamental critique of direct democracy.

The legal notion of legitimacy is closely connected to constitution-alism (Schliesky 2004: 164). According to modern understanding, the people themselves frame the constitution as an act of self-deter-mination. The people hold the *pouvoir constituant* (constituent power) (Böckenförde 2004: marginal n. 5). Many modern constitutions have emerged from societal upheavals, and can be understood as decisions or contracts, which channel political power and stabilize the commu-nity.[271] They derive their legitimacy from the ideological and socioeco-nomic contexts, the prospect of continuity and peaceful change, and the approval of the people (Badura 2003: Chapter A, marginal ns 7; 9).

The constitution is the binding set of rules for a given community (Hesse 1999: marginal n. 16). It sets out the universally accepted proce-dures on the transfer and exercise of authority: elections, lawmaking, government, and adjudication. Furthermore, it may contain the objec-tives of government (Badura 2005: Chapter A, marginal n. 7). Thus, the constitution becomes the measure for legitimacy in the modern constitutional state.

In this context, the distinction between legality and legitimacy becomes relevant. The former describes the conformity of any activ-ity with a predefined set of rules, whereas the latter stands for the jus-tification of government authority (Würtenberger 1982: 677–9; 711–5; Quaritsch 1987: marginal n. 1989). Legality is one of the most important aspects of modern statehood, as it limits and allows the effective con-trol of governmental authority (Würtenberger 1982: 679). Nevertheless, it is also embedded in the broader context of legitimacy. The rules that define the standard of legality and bind governmental authority must themselves be legitimate.

Whether legality is an entirely formal concept depends on the per-spective and the content of the constitution. From a legal postitivist perspective, which strictly separates law from moral imperatives, legal-ity, the observance of the constitutional requirements, is the same as legitimacy (Schliesky 2004: 167). From a natural law viewpoint, non-legal values like justice, rightness, or truth, complement legality, as cri-teria of legitimacy (Röhl 1987: 176).

The exact legal definition of legitimacy – whether it includes non-legal values or not – eventually depends on the constitution. A constitution

---

[271] For example, the origin of the Constitution of the United States is inextri-cably linked with the American Revolution and the dissociation of the colonies from British rule. In a similar way, the German *Grundgesetz* organized and stabi-lized political power in post-Fascist West Germany.

might contain only neutral principles with which to organize decision- and lawmaking processes or it might also provide material principles and objectives for government. In the latter case, values are part of the constitution, which must be taken into account in all governmental procedures. The inclusion of such material provisions provides a gateway for moral-ethical considerations (Schliesky 2004: 169).

The constitution, a universal framework, binds a community and the exercise of authority therein, making it the yardstick for legitimacy (Badura 2005: Chapter A, marginal n. 9). Therefore, as long as government authority – in particular the creation and application of law – complies with the provisions of the constitution, it conforms to the prerequisites of legality, thus making it legitimate. The relevant category in the assessment of legitimacy is the constitutionality of government acts. Whether this also satisfies the ethical and moral standards remains an issue.

Constitutionality

From a lawyer's perspective, legitimacy is considered a matter of constitutionality. Constitutional provisions stipulate the benchmark for all government activities. Therefore, by complying with these constitutional specifications, government actions are legitimate. Since the focus of the empirical part of this book has been on German administrative practice, this section will take into consideration the provisions of the German *Grundgesetz*.

Art. 20 para. 2 of the *Grundgesetz* is the starting point for any discussion of legitimacy in German constitutional law (Schliesky 2004: 232). According to this provision, all state authority is derived from the people. Thus, the provision recognizes the sovereignty of the people and establishes democracy as the prevalent legitimacy model of the *Grundgesetz*. The Federal Constitutional Court's second senate, and, in particular, its long-time member *Ernst-Wolfgang Böckenförde* formulated the dominant theory on the interpretation of Art. 20 para. 2 of the *Grundgesetz*. State authority within the meaning of this paragraph encompasses all state activity, and not just those measures that are usually considered as falling within the exclusive domain of the state. Consequently, "[a]ny official activity resulting in decisions is an exercise of state authority requiring legitimacy."[272]

---

[272] Federal Constitutional Court, Judgement of 26 June 1990: 73; 114–15 [translation by the author].

Democratic legitimacy can be provided through three mechanisms:

- Functional and institutional legitimacy
- Organizational-personal legitimacy
- Content-related legitimacy (Böckenförde 2004: marginal ns 14–25).

Functional and institutional legitimacy cover the aspect of the democratically legitimized exercise of state functions through specific organs and institutions. In the words of the Federal Constitutional Court, "[t]he legislative, executive, and judiciary organs derive their institutional and functional democratic legitimacy from the decision of the *pouvoir constituant* as stipulated in Art. 20 para. 1 of the *Grundgesetz*."[273] Legitimacy is also linked with the principle of the separation of powers.

Organizational-personal legitimacy is a necessary complement to functional and institutional legitimacy. The latter serves as the foundation for government organs. However, citizens not only create organs and institutions, they also decide who will take office. Therefore, the former constitutes an uninterrupted legitimacy chain, which runs from the people to the individual officeholder. The legitimacy chain does not necessarily have to connect each incumbent with citizens. Legitimacy can also be conveyed indirectly. Significant in this regard is the parliament, which is, as the main representative organ, the first and most important link in the legitimacy chain.

The purpose of organizational-personal legitimacy is not to provide for the legitimacy of the incumbent, but to ensure the legitimacy of all state actions. For mixed bodies, composed of persons that are both within and outside of the legitimacy chain, the majority of its members must be democratically legitimized. While it is useful to let those who represent societal interests participate in the decision-making process, decisions made by such organs must never depend on the non-legitimized members.[274] Böckenförde argues that civil society actors often lack democratic legitimacy. Although he sees a role for them in certain social decision making processes, he denies them any legitimating function (Böckenförde, 2004, marginal ns 29–30). In a similar vein, Schmitt Glaeser declared: "[societal] participation thus has no legitimizing function" (Schmitt Glaeser 1973: 220).

---

[273] Cf. Federal Constitutional Court, Order of 8 August 1978: 125; Cf. also Federal Constitutional Court, Judgement of 17 July 1984: 88.

[274] Federal Constitutional Court, Order of 24 May 1995: 67–8.

Content-related legitimacy means that the substance of policies and state activities – particularly the regulatory content of laws – must originate with the people. This is guaranteed through the political accountability of office holders. Members of the German Parliament are directly responsible, because they stand for election every four years, and are thus accountable to the electorate. The exercise of state authority is defined in Art. 20 para. 3 of the *Grundgesetz*: legislative power is vested in Parliament, a representative organ that is most directly connected to the people. All other bodies and state activities must conform to the laws enacted by Parliament in order to conform the will of the people.

All three forms of democratic legitimacy are intertwined and complement each other. None of these forms can maintain effective democratic legitimacy alone. Accordingly, the Federal Constitutional Court does not demand that a particular form of democratic legitimacy is observed under certain circumstances. It is imperative that a certain level of legitimacy (*Legitimationsniveau*) is maintained, one that effectively ensures that the exercise of authority has its origin in the people.[275]

The necessary level of legitimacy depends on the impact of the decision. An important constitutional element that becomes relevant in this context is the principle of the rule of law (*Rechtsstaatlichkeit*),[276] stipulated by Art. 20 para. 3 of the *Grundgesetz*. From this principle emanates the principle of subjection to the law (*Vorbehalt des Gesetzes*):[277] all state activities that affect the legal positions of individuals falling in the scope of basic rights must be legitimized by an Act of Parliament.[278] This means that the substantive aspects of measures impinging on basic rights must be regulated by Parliament.[279] This concept – termed *Wesentlichkeitstheorie*[280] – reinforces the legitimacy of laws, allowing courts to examine thoroughly the constitutionality of laws and return faulty legal provisions to the parliamentary lawmaking process (von Bogdandy 2000: 185). Democratic legitimacy is achieved if

---

[275] Federal Constitutional Court, Order of 24 May 1995: 67.

[276] Cf. Singh 2001: 11–12 on the German concept of rule of law from the common law perspective; cf. also Foster and Sule 2002: 163–4.

[277] For a brief discussion of the *Vorbehalt des Gesetzes* cf. Schmidt-Aßmann 2001: 307–8.

[278] Federal Constitutional Court, Order of 28 October 1975: 248–9; Federal Constitutional Court, Order of 8 August 1978: 126.

[279] Federal Constitutional Court, Order of 8 August 1978: 126–7.

[280] For a brief English-language discussion of this concept cf. Badura 2001: 51.

constitutional procedures are observed. Thus, constitutional legality equals democratic legitimacy.[281]

From a political science perspective, the concept of legitimacy developed by the Federal Constitutional Court and Böckenförde amounts to a strict version of input legitimacy (Schliesky 2004: 235). Additional mechanisms furthering legitimacy are deemed unconstitutional, because they infringe upon the basis of democracy – the freedom of elections and the unrestricted mandate of representatives.[282]

Of course it must be noted that this concept of legitimacy has been subject to intense criticism. For example, Rinken argues that the determination of the Federal Constitutional Court's concept as the only valid legal theory of legitimacy is an expression of judicial activism, which hampers diversity in the political process and restricts the leeway of political actors (Rinken 1996: 304ff.). Blanke finds fault with the Federal Constitutional Court's concept on the grounds that it stipulates a hierarchical, premodern model of administrative organization, restricting the erection of contemporary administrative structures (Blanke 1998: 4701). Fisahn expands on these positions when he criticizes the Federal Constitutional Court's concept as being a closed system, which shields the state and its institutions from empirical opinions and interests (Fisahn 2002: 230–1). Participation, he suggests, could counter the autonomization tendencies of the administration, as it might establish a supplementary control mechanism in addition to parliamentary control (Id.: 245).

---

[281] Federal Constitutional Court, Judgement of 25 January 1983: 47; Criticism of this notion is documented by Schliesky 2004: 248–53. Cf. also Brugger 2000: 60ff.

[282] Federal Constitutional Court, Judgement of 25 January 1983: 47.

15
Legitimacy and Law Beyond the State

Since globalization involves the coordination of policies beyond the national level, the question emerges regarding if and how such processes are democratically legitimized. It could be that transnational and supranational coordination and cooperation suffer from a lack of democratic legitimacy. Usually, states will negotiate on the coordination of their activities or agree to cooperate, because the benefits of concerted action exceed those of individual action. The effects on democracy are twofold. First, a basic condition of democracy is that the electorate has a choice. Intergovernmental or multilateral negotiations result in a final agreement, for which approval must be sought at the national level. The electorate (or Parliament, for that matter) has the choice only between approving and rejecting the agreement; alternatives regarding the content cannot be discussed. Second, as continuity is a key characteristic of foreign policy, it is unlikely that a change of government will lead to a revision of previous agreements. Here, again, the electorate is bereft of a choice (Scharpf 1993: 168–9).

The preceding discussion on legitimacy in the social sciences indicated that there are problems with legitimacy in contexts beyond the state level. For this reason, the following section will describe the legal view of legitimacy in contexts beyond the exclusive jurisdiction of the state. This discussion will proceed, first, by examining the situation within the supranational context of the EC and, second, by looking at the problem of legitimacy in both international and transnational contexts.

Legitimacy and supranational contexts

The Federal Constitutional Court's construction of democratic legitimacy was conceived in a national context. It faces grave problems in a supranational setting.

The EC may exercise authority over its member states and their citizens. One way this happens is through legislative acts. According to Art. 249 of the EC Treaty, the EC can enact regulations, which are directly applicable within the member states, and directives, which must be implemented by member states. Under this mechanism, the European legal order overarches the domestic legal systems of member states. The EC's own administration offers a second way to exercise authority.[283] Third, the European Court of Justice (ECJ) and the European Court of First Instance perform judiciary functions (Art. 220ff. of the EC Treaty).

Since the EC exercises authority over Germany and its citizens, the question of its legitimacy arises. According to the concept of legitimacy developed by the Federal Constitutional Court and Böckenförde, an uninterrupted legitimacy chain must run from the German people via Parliament to those organs and officeholders issuing legal acts that in the end affect the German people. A major problem in this regard is the length of the legitimacy chain. The Federal Constitutional Court requires that the legitimacy chain is not interrupted.[284] However, it makes no remarks on its length. Some authors have argued that the legitimacy chain may not only loosen its function if it is interrupted, or become weak. This objection is rejected as a "political value judgement" (Schmidt-Aßmann 1991: 360). But two examples demonstrate the impossibility of organizational-personal legitimacy in supranational settings.

The first example concerns the participation of the national parliament in the process of choosing Commissioners for the European Commission.

It is not entirely up to Germany's government to decide who will be its representative at the European Commission. The Council and the nominee for the President of the Commission collectively draw up a list of nominees for positions in the Commission in accordance with the proposals submitted by the member states. The Council may act by qualified majority (Art. 214 para. 2 sub. 2 of the EC Treaty). Subsequently, the nominees are collectively approved by the European Parliament, and are then appointed by the Council again by a qualified majority (Art. 214 para. 2 sub. 3 of the EC Treaty). Either the Parliament or the Council could approve or appoint a German candidate without

---

[283] Cf. for example the role of the European Chemicals Agency, Arts 71ff. of Regulation (EC) No. 1907/2006.

[284] Cf. Federal Constitutional Court, Order of 15 February 1978: 275; Federal Constitutional Court, Order of 24 May 1995: 67.

German consent, in which case, the legitimacy chain is broken. Even if a German candidate is approved and appointed with consent of the German members of the European Parliament and the German vote in the Council, the candidate is not personally legitimized by the German people or their representatives in the German Parliament (Schliesky 2004: 395).

Another example relates to the indirect implementation of European law, that is, the execution of European provisions via the national administration.[285]

In this case the organizational-personal legitimacy problem also emerges. The European Council and the European Parliament legislate these rules; but, the national administration is not accountable to the European legislator, but to the national parliament. The national parliament, however, has little or nothing to do with the enactment of European legal acts (Winter 2005: 268). Ideally, the legislator also oversees the process of the application of laws and holds the administration accountable. In the context of applying European legal acts, this clearly is not the case.

Apparently, the national practice requiring an undisrupted legitimacy chain cannot be met in the supranational context of the EC.[286] According to the Federal Constitutional Court, the loosening of the legitimacy chain in a supranational setting complies with the *Grundgesetz* and the requirements for democratic legitimacy contained therein. First, under Art. 23 para. 1 the *Grundgesetz* allows for the transferral of sovereign rights to a supranational entity. A supranational organization furnished with sovereign rights has authority over member states. The scope of authority depends on the scope of the delegated sovereign rights. But it is key to the exercise of authority that decisions made by the supranational organization bind its member states. This includes instances where a dissenting member state must respect majority decisions. Second, the political responsibility for the delegation of authority to a supranational organization rests with the *Bundestag* as the immediate representation of the people. Approving the law that sets out Germany's accession to the supranational organization constitutes the necessary act of democratic legitimacy.[287]

---

[285] Cf. Oppermann 2005: §7, marginal ns 31ff. for details on the implementation of Community law.

[286] Federal Constitutional Court, Judgement of 12 October 1993: 182; Cf. also Schliesky 2004: 396; Grams 1998: 129.

[287] Federal Constitutional Court, Judgement of 12 October 1993: 183–4.

One consequence of this is that the German people are not the exclusive point of origin for the authority applied within German territory which impacts German citizens (Schliesky 2004: 399). In a way, the Federal Constitutional Court has departed from its input-oriented concept of legitimacy, taking into account aspects of output legitimacy. In this case, the outcome – European integration and particularly securing confidence in the European currency[288] – serves as a legitimizing factor.

The question still stands as to how authority in supranational or multidimensional settings can be secured. In such contexts, a pluralist legitimacy model made up of several elements should be considered. There are input-related mechanisms, which help to legitimize the organs of the EU and the EC. The European Parliament – elected by the citizens of the member states – participates in the lawmaking process and controls other Community organs. However, the European Parliament merely represents the peoples of the member states, not the European people as a whole. Since the last enlargement of the EU and the EC, its inhabitants are more heterogeneous than ever. The cultural, ethnical, linguistic, political, economical, religious, and ideological variety hampers the emergence of a European demos (Ossenbühl 1993: 634; Graf Kielmansegg 1996: 55).

Input legitimacy in the EC has deficits.[289] Therefore, it must be complemented by elements of output legitimacy, since input legitimacy cannot be sufficient if authority is exercised on a supranational level with the explicit purpose of performing certain tasks and achieving predefined goals. The purpose and aims of exercising authority must be included in considerations of legitimacy (Schliesky 2004: 661). Relevant factors for the determination of output legitimacy are the efficiency and efficacy of the exercise of authority when performing the functions assigned to the state. However, with the introduction of these factors as legitimizing elements of supranational authority the question arises as to how to output legitimacy can be measured. The purpose and goals of authority are criteria for determining the efficacy and efficiency of authority. Evaluation must occur before and after any exercise of authority (Id.: 670ff.; Peters 2001: 580ff.). Evaluating the outcome of certain measures *ex ante* is a common function of courts and legislative or administrative bodies. Courts, for example, assess the presumed outcome of a measure when issuing a preliminary

---

[288] Federal Constitutional Court, Judgement of 12 October 1993: 208–9.

[289] For an overview of the EU's democratic deficiencies cf. Grams 1998: 93ff.

injunction.[290] Another example is the assessment of a law's impact as an element of the lawmaking process. Practical criteria for an *ex post* legal assessment do not exist as yet. Here, the law and social sciences must develop proper methods (Schliesky 2004: 672).

Peters formulates a concept of *ex post* assessments against the backdrop of a constitution for Europe. According to her approach, a European constitution would be justified if it contributed to the welfare of the people, that is, legitimacy through proof of practicability (*Legitimation durch Bewährung*) (Peters 2001: 580). In more abstract terms, this approach ties in with the general purposes of the state. The mainstay of this legitimacy concept is the efficacy of state measures – what is legitimate is what best fulfils the various purposes of the state. This, however, requires the installation of measures to control the success of legislative acts – which has yet to happen (Grams 1998: 343–4).

Legitimacy and foreign relations power

After applying insights gleaned from Federal Constitutional Court and other academic approaches to legitimacy strategies in a supranational context, the problem of legitimacy and foreign relations power (*Auswärtige Gewalt*) can be explored.

The legitimacy problem regarding the exercise of foreign relations power presents itself in a slightly different way than in the supranational context sketched above. Here, the exercise of this power occurs at the international level. Because of the equality of states that emanates from their sovereignty, actors can be viewed as being at the same level. A supranational structure comparable to the EC, which binds its members together, does not exist at the international level. Organizations with the power to issue legally binding decisions are scarce. If an IO has been vested with this power, such decisions usually rest on a consensus, giving objecting states the power of veto.[291] Nevertheless, IOs yield authority that has an impact on the policy creation and lawmaking in member states (Delbrück 2003: 35).[292]

---

[290] Cf. §123 of the *Verwaltungsgerichtsordnung* (Administrative Court Procedures Code) and §32 of the *Bundesverfassungsgerichtsgesetz* (Federal Constitutional Court Act).

[291] Cf. Art. 5 of the OECD Convention.

[292] Cf. Hobe 1998a: 294–308 on the impact of lawmaking by the WHO, FAO, and other IOs on the domestic legal order.

As previously demonstrated, this is also the case with transnational bureaucracy networks. In fact, the lawmaking processes of IOs originally lay in these networks.

Foreign relations power is a description of the competences of state organs in foreign relations as set out in the constitution. The *Grundgesetz* does not explicitly mention such a concept, but constitutional law scholars in Germany have operated with it for the last 100 years.[293] The concept of foreign relations power is rather broad, and follows from the various constitutional provisions relating to foreign affairs. Since it is not restricted to activities relevant to international law, for example, the conclusion or termination of international treaties, the concept is very broad. Activities with political or nonlegal implications also fall within the scope of foreign relations power (Grewe 1988: marginal n. 1; Wolfrum 1997: 39). In the age of globalization, where interior and foreign policies are so entangled that it is hard to differentiate between the two, the latter aspect of the foreign relations power gives rise to the question of whether the area covered by foreign policy can be concretely defined. A distinction can be made using the criterion of immediacy: any act that immediately affects a foreign actor or takes immediate effect outside a state's own territory lies within the field of foreign policy (Grewe 1988: marginal n. 5).

On the basis of Federal Constitutional Court rulings and the opinions of legal scholars, the exercise of foreign relations power falls to the executive. This follows from a comprehensive survey of all relevant constitutional provisions (Grewe 1988: marginal ns 40–51; Kokott 1996 937–8; Kommers 1997: 148ff.). Art. 65 of the *Grundgesetz* furnishes the Chancellor, the head of government, with the competency to determine the guidelines of government policy, including foreign policy. Parliament's role in foreign affairs is laid down in Art. 59 para. 2 of the *Grundgesetz*: political treaties require the consent or participation of Parliament. It follows from this provision that there must be international agreements that can be concluded by the government without seeking the immediate approval of Parliament. The Federal Constitutional Court construed these provisions to the effect that Parliament's role is restricted to cases concerning political treaties and treaties that have to be implemented through a federal act. It defines the former group of treaties as "agreements affecting the existence of the state, its territorial

---

[293] For an overview of the theoretical background and historical development cf. Grewe 1988: marginal ns 9–31. The term was first used by Haenel 1892: 531ff.

integrity, independence, or position in relation to the community of states." In particular, treaties that aim towards "maintaining, solidifying, or expanding the position of power of a state compared to other states" are political treaties within the meaning of Art. 59 para. 2 of the *Grundgesetz*.[294] The latter are treaties that concern subject matter that can be implemented only through acts of the federal government.[295] It emanates from the aforementioned *Wesentlichkeitstheorie* that this is particularly the case if basic rights are substantially affected (Geiger 2002: 135).

Apart from these cases, the exercise of foreign relations power is within the executive's domain. The Federal Constitutional Court explains that "[i]n parliamentary democracies the parliament is generally tasked with legislation and the executive performs governmental and administrative functions." Beyond parliament's participation in treatymaking, it is reduced to its "general constitutional control rights."[296] Accordingly, Art. 59 para. 2 s. 2 of the *Grundgesetz* provides that the government can conclude administrative agreements without seeking the approval of Parliament. This includes international treaties, which do not concern political relations or require implementation through a federal act.

The prominent role of government and the restriction of Parliament in this area stems from the idea that the government has the necessary flexibility and resources to deal with the complex problems that arise in foreign affairs.[297] According to the Federal Constitutional Court, this does not collide with the constitutional concept of legitimacy. The chain of legitimacy links Parliament with government, sufficiently legitimizing the executive and allowing Parliament to exercise its control function.[298]

In its various rulings on foreign deployments of armed forces, the Federal Constitutional Court has not deviated from the rationale of its former decisions. According to the Court, the requirement of Parliament's approval prior to the foreign deployment of armed forces follows from the special constitutional provisions on national defence.[299]

---

[294] Federal Constitutional Court, Judgement of 29 July 1952: 381–2 [translation by the author].

[295] Federal Constitutional Court, Judgement of 29 July 1952: 381.

[296] Federal Constitutional Court, Judgement of 29 July 1952: 394 [translation by the author].

[297] Federal Constitutional Court, Judgement of 17 July 1984: 89.

[298] Federal Constitutional Court, Judgement of 17 July 1984: 88 and 109.

[299] Federal Constitutional Court, Judgement of 19 and 20 April 1994: 381; Order of 25 March 1999: 269; Federal Constitutional Court Order of 25 March 2003: 44.

The restricted role of Parliament is deemed problematic in a globalizing world, where virtually every facet of life is affected by events occurring beyond national boundaries (Hailbronner 1997: 11). The vast majority of Germany's international obligations stem from administrative agreements within the meaning of Art. 59 para. 2 s. 2 of the *Grundgesetz* (Kadelbach and Guntermann 2001: 565). These came about without the participation of Parliament. But even if an international treaty has to be approved by Parliament, because of its nature, Parliament has only the choice of consenting or rejecting it. §82 para. 2 of the *Geschäftsordnung des Deutschen Bundestag* (Rules of Procedure for the German *Bundestag*) stipulates that motions to amend treaties with foreign states are not admissible. Since the political costs of rejecting such an agreement are great, in actual fact, Parliament has little opportunity to influence and shape such political processes.[300]

The limited capacities of Parliament to engage in the exercise of foreign relations power are further illustrated by looking at the role of IOs. In view of the importance of IOs, which provide a solid platform for interstate coordination and cooperation, to the international system it is not surprising that founding an IO or accedding to one requires the consent of Parliament. Accordingly, Art. 24 para. 1 of the *Grundgesetz* allows the transfer of sovereign rights to IOs. The transfer must occur only with the express approval of the legislative branch.[301] Thus, Parliament decides whether Germany will seek membership of an IO. Nevertheless, Germany has joined the WHO and FAO by means of an administrative agreement without seeking parliamentary consent.[302] However, Parliament did consent to Germany's accession to the OECD.[303]

Under Art. 23 para. 3 of the *Grundgesetz*, Parliament may become involved in the decisionmaking processes of supranational organizations, but it does not have powers to review the work of an IO as long as the IO acts within the limits of the treaty that governs it (Kadelbach 2003: 49). If the IO acts within the aims of its founding treaty, it merely makes use of the powers conferred to it. It is only when these measures

---

[300] The German Parliament is aware of this problem, cf. *Deutscher Bundestag, Schlussbericht der Enquete-Kommission, (BT-Ds. 14/9200)*: 445–6.

[301] Cf. *supra* 19.

[302] Cf. *Bekanntmachung der Satzung der Ernährungs- und Landwirtschaftsorganis ation der Vereinten Nationen; Bekanntmachung der Satzung der Weltgesundheitsorganis ation*. Cf. also Kadelbach 2003: 42–6.

[303] *Gesetz zum Übereinkommen vom 14. Dezember 1960 über die Organisation für Wirtschaftliche Zusammenarbeit und Entwicklung (OECD)*.

amount to changes to the founding treaty that parliamentary approval is required.[304]

If an IO issues legally binding decisions to be implemented into national law, Parliament is involved in the incorporation, since it has to pass the necessary law (Wolfrum 1997: 52). However, the incorporation can only be accepted in full or rejected. If the IO's resolutions are nonbinding, they are soft law. Despite its significance in international affairs, soft law is not regarded as a political treaty within the meaning of Art. 59 para. 2 of the *Grundgesetz*, and thus does not require parliamentary approval (Kadelbach 2003: 50).

The current constitutional arrangement assigns Parliament a rather passive role in the exercise of foreign relations power. In the case of soft law, in particular, it can only make use of its control function, but can not actively participate in legislative processes. The principle of division of powers allows only for *ex post* parliamentary control, so Parliament may not engage in ongoing decisionmaking in the foreign relations arena.[305]

Parliament may employ its right of control to request information from the government on foreign affairs. The government, however, is not obliged to inform Parliament without a prior request. Parliament has various tools at its disposal when exercising its control rights.[306] First, according to Art. 43 para. 1 of the *Grundgesetz*, Parliament itself or one of its committees, for example, the Committee on Foreign Affairs, may summon members of the government – the Chancellor and ministers – and thus require them to be present during sessions. This gives Parliament and its committees also the right to submit questions to the member of government.[307] In addition, the Committee on Foreign Affairs influences Germany's foreign policy by conferring with government officials. This allows the government, and, in particular, the minister of foreign affairs, to determine whether their current foreign policy stance has the support of Parliament (Münzing and Pilz 1998: 601). Another available tool is the Parliament's right to submit inquiries to the government ("interpellation"). If the government refuses to reply, the matter may be discussed in Parliament (Achterberg 1984: 467ff.).

---

[304] Judgement of 22 November 2001: 1560–1.
[305] Federal Constitutional Court, Judgement of 17 July 1984: 139.
[306] Federal Constitutional Court, Judgement of 17 July 1984: 109–10.
[307] Cf. Badura 2005: Chapter E, marginal ns 17 and 46 for details.

In practice, Parliament makes good use of these instruments. In recent years, a large number of parliamentary debates have been dedicated to foreign policy, and many inquiries have been conducted into the government's foreign policy. As the *Bundestag* operates as an *Arbeitsparlament* ("working parliament"),[308] its committees deal with various matters of foreign policy (Krause 1998: 144ff.).

Since lawmaking processes increasingly occur beyond state lines and have their roots in informal structures, the question arises as to whether Parliament's *ex post* control rights are sufficient to legitimize such activities. The Federal Constitutional Court has stated that the legitimacy chain runs from Parliament to the government, legitimizing the government in its exercise of the foreign relations power. However, this notion can be disputed. The first objection, already raised in the preceding discussion of the EC, concerns the length of the legitimacy chain. The chain runs from citizens eventually to the German delegate acting on behalf of Germany – via Parliament via the Chancellor via the minister via the ministerial staff, etc. In this case one might argue that the legitimacy chain is too stretched to be effective. The fact that Parliament approved the accession treaty means that it is, at the very least, aware of the length of the legitimacy chain. However, where the Germany's membership in an IO is the result of an administrative agreement within the meaning of Art. 59 para. 2 s. 2 of the *Grundgesetz*, Parliament's involvement in this process is insufficient. In this case, the existence of an effective legitimacy chain is doubtful.

Second, as is the case with Germany's participation in the EC, decisionmaking occurs in mixed committees. German delegates, who have not been approved the German people, sit with delegates from other nations. The Federal Constitutional Court has established the requirement that democratically legitimized personnel must form the majority in mixed committees. In international or transnational contexts, this requirement cannot be fulfilled.

In view of this deficit of democratic legitimacy, other methods have been proposed to ensure the legitimacy of international acts that have repercussions at the national level. There are three possible approaches. The first and second approaches tie in with the concept of input

---

[308] The archetype of the "working" parliament is the US Congress with its large number of specialized committees and subcommittees on various policy matters. The opposite is a *Redeparlament*, a "talking" parliament. The United Kingdom's House of Commons embodies the archetype of "talking" parliament – a parliament in the original sense of the word.

legitimacy, and focus on the role of Parliament and its representatives. The third centres on the concept of output legitimacy, relying on alternative legitimation strategies.

The first approach seeks out domestic solutions to the democracy deficit in foreign affairs. Accordingly, Parliament's role in foreign affairs could be strengthened, mimicking its involvement in European affairs. Here, Art. 23 para. 3 of the *Grundgesetz* provides for close cooperation between the government and Parliament. Parliament – particularly the Committee on Foreign Affairs – could also be furnished with similar rights (Kadelbach 1992: 54–5). The obligations of the government to brief Parliament on the state of affairs and seek out Parliament's view on foreign policy matters could be regulated by a specific law, similar to that regulating cooperation between the government and Parliament in European affairs or the involvement of Parliament in the decision to deploy armed forces abroad.

The second approach, proposed to ensure the legitimacy of international acts, is based on international solutions. Here, the focus is not on the acts of an individual national government. Instead of legitimization via national parliaments, international or supranational parliaments might help to legitimize international authority. Models for this approach are the parliamentary assemblies existing within the OSCE framework or the Council of Europe, which contribute to the legitimacy of an IO's acts (Marschall 2002: 385–6; Slaughter 2004: 108ff).[309]

Various proposals have been advanced for the creation of a UN Parliamentary Assembly (Czempiel 1995: 42; Held 1995: 108ff.; Archibugi 1995: 135ff.).[310] However, the reform of the UN's organizational structure has yet to happen. Furthermore, such a measure would raise the general question as to how democratic governance on the international level can be accomplished. The prevalent notion is that democracy is achieved via the principle of equality among states, which flows from their sovereignty: one state, one vote. However, democracy literally translates into "rule by the people." It is the focus on the individual that defines democracy. On the domestic level this means: one person, one vote. This notion stands in stark contrast

---

[309] Cf. Ipsen 2004: §34, marginal n. 14 for an overview of these parliamentary assemblies.

[310] European Parliament resolution on the reform of the United Nations, para. 39; *Deutscher Bundestag, Für eine parlamentarische Mitwirkung im System der Vereinten Nationen (BT-Ds.15/5690); Deutscher Bundestag, Für eine parlamentarische Dimension im System der Vereinten Nationen (BT-Ds. 15/3711).*

to the first. Tuvalu, the UN member state with the smallest population, has roughly 12,000 citizens, whereas China has 1.3 billion. The
discrepancy is obvious. According to the first concept of democracy,
Tuvalu has one vote in the UNGA, just like China. The second view
would allot China with more than 100,000 times the votes of Tuvalu
(Bodansky 1999: 614). To avoid the translation of a country's population into votes, some scholars pursue the idea of an international
legislative body, elected by the people of the world. However, the
most significant objection is that the necessary global demos does not
presently exist in the world today. A demos shares common values,
culture, and experiences that evoke a sense of community and identification with the political system. Otherwise, majority rule will not
be accepted (Scharpf 1993: 165–6).[311] However, points of intercultural
connection through intercontinental air transportation or the Internet
largely exist in industrialized states only; and intercultural connec-
tions do not necessarily bring about cultural acceptance. In addition,
while a democratic culture exists in industrialized states, can there be
such a thing as state representatives to a world parliament, if global
citizens only know democracy as a vague concept? The people of the
world share the same planet and face the same problems; but this does
not make them the tight-knit community, or demos (Delbrück 2003:
36–7; Scholte 2002: 292). As a result, the chance for a world parliament
acting as a legitimizing source is slim. Furthermore, a world parliament cannot legitimize individual national governments acting at the
international level.

Lack of parliamentary control of activities beyond exclusive national
jurisdiction is further aggravated by the fact that many activities occur
in informal settings. The empirical part of this book showed that transnational bureaucracy networks carry out a host of activities – including
formulating standards that later become transnational public law. It has
been pointed out that such types of networks are generally regarded
to have a legitimacy deficit, mainly due to their informal, obscure
nature.[312] The question is of whether parliaments can contribute to mit-
igate the legitimacy deficit. The problem is that parliamentary control
depends on the existence of formal structures (Marschall 2002: 390),
making it difficult to control transnational bureaucracy networks.

---

[311] Cf. also Claude 1960: 133ff. for an account of the practice of "Veto"- in
the UN, an example of the non-acceptance by the "Big Five" of majority rule;
Böckenförde 2004: 63ff.

[312] *Supra* 42.

This problem had been recognized as early as the 1970s, after Keohane and Nye first pointed out the emergence of transnational relations. According to Kaiser, such interactions – "[t]he intermeshing of decisionsmaking across national frontiers and the growing multinationalization of formerly domestic issues..." – pose a threat to democracy (Kaiser 1971: 706). He feared that the outcome of such interactions might result in a technocratic rule, devoid of the primacy of politics (Id.: 717). Foreign policy is already an area that escapes the full scrutiny of Parliament. With the increasing globalization of regulatory matters and the enmeshment of bureaucracies, certain subject areas that normally fell within national jurisdiction are now becoming internationalized, and becoming a matter of foreign policy, so the political weight shifts from the legislative to the executive (Id.: 710–13). He suggested that this development be countered by including parliamentary representatives on transnational committees or by creating councils of elected experts to oversee such activities. Additionally, the administration would have to adapt and implement specific control measures (Id.: 718–19).

Slaughter also suggests that legislators form networks of their own ("legislative networks") (Slaughter 2004: 112ff.). Such networks have emerged in the past few years, for example, the Parliamentary Network of the World Bank. Networks of this type can serve as a quasidemocratic corrective for technocratic developments (Id.: 119). Despite these developments, however, in comparison to networks of administrative officials, legislative networks still lag behind (Id.: 127).

The third approach, which largely relates to output legitimacy in European affairs, is the development and application of alternative legitimacy strategies that build on elements of democratic legitimacy, such as transparency and accountability, but can also function independently. Other such elements include efficiency and expertise (Delbrück 2003: 34).

Increasing the visibility of the network structures improves their transparency and also makes networks more accountable (Id.: 43). It is also suggested that information on network activities is made available at a central place, such as the Internet. Some have rejected this approach on the basis that it will inevitably lead to information overload (Slaughter 2004: 235). It is, however, one way in which interested citizens can obtain information on transnational activities.

NGOs also contribute to the legitimacy of lawmaking activities that occur beyond the exclusive jurisdiction of the nation state. Their role is twofold. First, they provide input legitimacy, as they can serve as conveyors for the will of at least a fraction of those governed. They contribute to a "pluralistic international public discourse" (Delbrück 2003: 41). They also can give a voice to stakeholders who might otherwise be underrepresented, because they lack resources, power, or knowledge to gain public attention (Scholte 2002: 293). Second, NGOs can help to support elements of output legitimacy. Through participation in international or transnational public governance, even as observers, NGOs can strengthen the transparency of the lawmaking process (Id.: 294; Slaughter 2004: 220). Transparency means that decisionmaking structures are revealed. In the case of transnational bureaucracy networks, NGOs not only represent civil society interests within the network, but can also carry information on the network's activities back to civil society.

The major objection to the NGOs' role in legitimizing international lawmaking is that they lack democratic legitimacy. Their own structure is not always democratic, and they generally do not represent majority interests. On a practical level, they can provide otherwise closed structures with an outside perspective and assist in publicizing the activities of IOs and transnational bureaucracy networks (Delbrück 2003: 41–2).

Conclusion

What can be learned from these examples of the legitimization of activities outside of the jurisdiction of the nation state? Decisions taken at the international and transnational level clearly lack legitimacy. This deficit can not be easily mended by strengthening the role of the national parliament in international affairs. It is also apparent that the Federal Constitutional Court's requirement of an undisrupted legitimacy chain, which relies exclusively on the concept of input legitimacy, is inadequate in international and transnational settings. The traditional, time-tested legitimacy strategies developed in regards to the nation state cannot simply be transferred to the international or transnational level (Delbrück 2003: 30). Parliamentary oversight of government or administrative activities in international or transnational settings is hard to accomplish, mainly because such activities lack visibility and are largely informal. Nevertheless, scholars suggest that Parliament should be given a greater role in this area, either by including parliamentarians in delegations of administrative officials or through the creation of parliamentary networks that parallel administrative network structures. Another approach aims at installing parliaments at the specific level where authority is exercised. While a European Parliament has been set up to represent the peoples of Europe, the establishment of a similar institution at the international level has not matured beyond the state of discussion and faces massive objections.

Other legitimizing mechanisms also come into consideration here. The model for legitimacy is still democratic legitimacy, where the will of the people is mediated by Parliament. However, this model is hard to implement in its traditional form in international and transnational contexts; legitimacy has to be accomplished by employing elements other than simple input legitimacy. Here, procedural elements like transparency and public participation come into play, and output legitimacy also becomes relevant. Hence, the effectiveness, efficiency or simply the practicability of measures can serve as criteria for their legitimacy. Although, upon initial examination, these elements can contribute to the legitimacy of a measure, some aspects remain unresolved. Can an NGO successfully and legitimately represent the views of civil society if it lacks its own democratic structure? How should effectiveness, efficiency, and practicability be assessed?

Two things become clear after studying legitimacy in contexts beyond the nation state. First, there is still a great deal of uncertainty among legal scholars on how to devise a feasible way of ensuring the legitimacy of activities occurring outside of the exclusive jurisdiction of the nation state. Second, there is no "silver bullet," that is, there will be no single element that exclusively conveys legitimacy. Instead, several elements will have to come together to be an adequate substitute for the input-centred democratic legitimacy of the nation state.

16
Legitimacy and Technical Standards

Technical standards are often created despite a certain degree of scientific uncertainty. This uncertainty often makes it difficult to determine what is scientifically necessary or technologically feasible. Accordingly, it must be possible to adapt technical standards quickly as scientific understanding grows. Organizations with a role in developing such standards must command the necessary resources to improve scientific understanding and be flexible enough to allow for future change (Majone 1984: 19–20).

The state's task of protecting human health and the environment necessitates the creation of state institutions that develop technical standards to regulate these areas. However, when acting within its discretionary authority in this capacity, does the administration run the risk of becoming too autonomous, thereby evading governmental control? The following section will illustrate this problem by looking at standard setting in Germany and in the EC.

One problem is that technical standards are influenced or created by specialized committees that operate outside the system of democratically legitimized actors. Nevertheless, these standards are often transposed into the legal system, becoming legally binding. Because Parliament is only minimally involved in the creation of technical standards, these often lack democratic legitimacy. Therefore, the establishment of technical standards outside Parliament and their inclusion in the legal order create a strained relationship between the need for scientific-technical expertise and democratic legitimacy. In such cases, there is a greater role for more output-related, legitimacy mechanisms.

Technical standards and the state

The first step is to clarify the reasons why technical standards must be legitimized. Two aspects should be mentioned in this regard.

The first concerns the context in which technical standards are used. Technical standards serve several purposes. They help to guarantee quality assurance and compatibility of goods – an electric plug must fit into a socket. And they ensure these goods are safe – the plug fits the socket without leakage of current or sparks. In general, pursuing these goals through standardization does not require legitimacy. Scientific organizations, economic associations, and other actors are free to set their own standards according to their own interests and aims. Since such standards are not legally binding, anyone can simply employ a different method to create a new standard if other standards appear inadequate.[313] Upon their incorporation into the legal order, technical standards become legally binding and consequently limit the discretion of individuals to apply their own standards. As technical standards thus in practical terms attain the status of a law, their legitimacy becomes an important issue (Gusy 1995: 106).

The second step concerns the content of technical standards. Standards relating to health or environmental safety inevitably have a political component, the product of weighing political, ethical, and economic values.[314] Limit values draw the line between safe and unsafe standardized test methods and help to determine the significance of data. Transportation guidelines delineate safe and dangerous modes of shipping (Lübbe-Wolf 1991: 235–6). The political nature of such decisions stems from their connection to basic rights such as health, protected under Art. 2 para. 2 s. 1 of the *Grundgesetz*, or from their connection to national policy objectives like the protection of the environment and animal protection, which are set forth in Art. 20a of the *Grundgesetz*. In Germany, the constitution requires that state measures bearing on basic rights be regulated by Parliament. Accordingly, legitimacy becomes an issue, especially if the standard in question encroaches on protected legal positions.

There are many ways in which technical standards can find their way into the legal order. One obvious ways is through the enactment of technical standards by parliamentary law. In Germany, for example, this has happened with leaded fuel. Parliament passed a law setting out the acceptable amount of lead compounds in gasoline.[315] While the legitimacy of technical standards enacted by Parliament is unproblematic overall, this approach is only useful in specific cases. Its flexibility

---

[313] However, the pressure to use these standards can be immense if the market leaders do so, cf. Gusy 1986: 248–9.

[314] For the case of limit values cf. Winter 1986: 14–15; Majone 1984: 21.

[315] §2 of the *Benzinbleigesetz* (Leaded Fuel Act).

is limited, since Parliament cannot respond to every scientific finding that would require an adaptation of the standard. Furthermore, Parliament generally lacks the expert knowledge to develop technical standards (Lübbe-Wolf 1991: 242).

As a result, the executive often determines technical standards. This often occurs through the government – usually the relevant minister – issuing an ordinance. For example, §23 para. 1 s. 1 *Bundes-Immissionsschutzgesetz* (Federal Control of Pollution Act) authorizes the federal government to issue ordinances on specified regulatory aspects. This sanctioned the issuance of certain ordinances such as the *Verordnung über elektromagnetische Felder* (Ordinance on Electromagnetic Fields), stipulating emission limit values for the intensity of electric and magnetic fields of high frequency installations. Parliament defines certain parameters for the creation of technical standards, the participation of stakeholder and experts, for example, and leaves the details to the government. This is constitutional under the principle of separation of powers, if Parliament defines content, purpose, and scope of the ordinance.[316] The government and its administrative apparatus have the resources to develop the relevant standards and revise them in a timely manner if scientific findings warrant amendments. While this satisfies the need for expertise and flexibility, Parliament can still maintain a high degree of control over the lawmaking:

- by virtue of its legitimacy as the organ directly elected by the people, it is its original task to control all activities of the legislature;
- by tailoring the authorizing law – within the constitutional limits – according to its own views of appropriateness;
- by reserving the power to override the ordinance.

Another possibility is the implementation of technical standards by means of administrative regulations (*Verwaltungsvorschriften*), which are issued to construe uniformly and concretize indefinite legal conceptions (*unbestimmte Rechtsbegriffe*)[317] such as "best available technique,"

---

[316] Cf. Art. 80 para. 1 of the *Grundgesetz*. For a discussion of delegation of legislation from a common law perspective cf. Singh 2001: 41ff. For the comparable intelligible principle under US constitutional law cf. United States Supreme Court, Mistretta v. United States, 488 U.S. 361 (1989).

[317] Cf. Singh 2001: 176ff. for an overview of this concept from the common law perspective. A survey of literature on comparative law offers a number of translations for the term *unbestimmter Rechtsbegriff*, for example "open-ended legal terms" or "indefinite legal terms." In the rest of this book, the term suggested by Singh – indefinite legal conception – will be used.

which Parliament frequently uses in laws with a scientific context.[318] The Federal Constitutional Court has ruled that Parliament's use of general principles in creating such administrative regulations is an effective way to safeguard basic rights. Instead of prescribing a fixed standard, the provisions are designed to be dynamic, allowing for the quick implementation of technical and scientific progress.[319] Therefore, the administration uses administrative regulations to concretize such provisions in the course of implementing the act. Usually, administrative regulations only govern the administration's internal conduct; but those that codify general principles inevitably have external effects.[320] Examples of administrative regulations that commonly employ this technique are those on noise[321] and air,[322] based on §48 of the *Bundes-Immissionsschutzgesetz*. These regulations place the government under an obligation to issue administrative regulations for the implementation of the act. Parliament determines the general framework for the standard, although it does not have to have regard for the constitutional requirements on ordinances. Administrative regulations based on §48 of the *Bundes-Immissionsschutzgesetz* require the consent of the *Bundesrat*[323], the Federal Council – a legislative organ at the federal level – to become law. The administration is assigned the task of developing from the general provisions in the act, because this job would overburden the Parliament's capabilities and competences. Here, the administration is better able to react to technical and scientific changes in a rapid and flexible way.

Occasionally, the executive sets up special committees to develop technical standards concretizing indefinite legal conceptions set out in the act. The composition of these committees is often mixed, and includes experts of various backgrounds. For example, the Committee for Hazardous Substances (*Ausschuss für Gefahrstoffe* – AGS), which was set up as an advisory body to the Ministry of Labour and Social Affairs, includes experts from employee associations, trades unions and other

---

[318] Cf. §3 No. 6 and §5 para. 1 No. 2 of the *Bundes-Immissionsschutzgesetz*.

[319] Federal Constitutional Court, Order of 8 August 1978: 137ff.

[320] Cf. Federal Administrative Court, Judgement of 19 December 1985: 320–1; Federal Administrative Court, Order of 21 March 1996: 523; Cf. also Lübbe-Wolf 1991: 224; Jarass 1999: 108–9.

[321] *Technische Anleitung zum Schutz gegen Lärm* (Technical Instructions on Noise Control).

[322] *Technische Anleitung zur Reinhaltung der Luft* (Technical Instructions on Air Quality Control).

[323] Cf. Arts 50ff. of the *Grundgesetz* on the *Bundesrat*.

relevant institutions. The members of the committee are appointed by the ministry[324] to elaborate upon Technical Rules on Hazardous Substances,[325] which define the best available technique in hazardous chemicals management.

Unlike the Committee for Hazardous Substances, which was established by an ordinance, the Nuclear Safety Standards Commission (*Kerntechnischer Ausschuss* – KTA)[326] came into being through an announcement by the Federal Ministry for the Environment, Nature Conservation, and Nuclear Safety in the *Federal Gazette*.[327] The announcement sets out the Committee's mission, and establishes its operating procedures and composition. Its task is the development of technical standards that define the best available technique in nuclear safety. These standards are not legally binding, but nevertheless they have been recognized in nuclear approval procedures.

Although, in all of these cases, the technical standards in question originate from institutions operated or controlled by the state, their legitimacy remains in question. Standards enacted by Parliament, or issued as ordinances by the executive, having parliamentary authorization and clearly defined conditions, are indeed democratically legitimized. However, the legitimacy of administrative regulations can be contested on the grounds that members of the executive branch rather than Parliament make substantial decisions on the basic rights of citizens. An additional problem regarding the legitimacy of administrative regulations arises when mixed committees like *AGS* or *KTA*, which are composed of persons lacking the necessary democratic legitimacy, issue them (Denninger 1990: marginal n. 144).

Another interesting development with respect to the legitimacy of technical standards in transnational public law is the incorporation of technical standards created outside the state's institutional structure into the domestic legal order. There are a number of reasons for the state in the end to rely on private standard-setting;[328] the most important is the lessening of the burden on state institutions. A strong example of this is the *Technische Anleitung zur Reinhaltung der Luft* (Technical Instructions on Air Quality Control), a technical standard adopted by the German administration. Many provisions of the *Technische*

---

[324] Cf. §21 of *Gefahrstoffverordnung*.

[325] *Technische Regeln für Gefahrstoffe*.

[326] For a critical review of the *KTA* cf. Denninger 1990: marginal ns 76, 131.

[327] *Bekanntmachung über die Neufassung der Bekanntmachung über die Bildung eines Kerntechnischen Ausschusses*.

[328] Cf. the list on Falke 2000: 248.

Anleitung zur Reinhaltung der Luft refer to standards set by private institutions (Mengel 2004: 128). This example emphasizes the importance of private technical standards and shows that since it is not possible for administrations to develop technical standards in all areas, they need to rely on private measures in some cases.

Private technical standards are incorporated into domestic law through either fixed reference (*starre Verweisung*), dynamic reference (*dynamische Verweisung*), or concretizing reference (*normkonkretisierende Verweisung*).

Fixed references point to a particular technical standard – specified in the referring provision through indications to its version and publication. The content of the technical standard referred to gains the same legal quality as the referring provision (Falke 2000: 249; Denninger 1990: marginal n. 137).[329] Dynamic references point to technical standards in their current version. This method is deemed to be very problematic from a constitutional perspective, as the legislator abandons its own lawmaking powers and issues a blank cheque to the private standardization organization. While in the case of the fixed reference the legislator is aware of the content of the standard, dynamic references indicate an intention not to review the standard in the near future, at least. Thus, the standard might change and eventually contain unconstitutional provisions without the legislator noticing (Denninger 1990: marginal ns 138–141).[330] The third method – concretizing references – ties in with the afore-mentioned indefinite legal conceptions. The legislator provides that a particular measure or project has to comply with the "best available technique" – it must be state of the art. This abstract objective is specified with the help of private technical standards, which eventually define the best available technique (Falke 2000: 252–3).

A host of private standardization organizations exists worldwide. The central standardization organization in Germany is called the *Deutsche Institut für Normung* (the German Institute for Standardization, DIN).[331] An agreement governs relations between the state and DIN.[332] The

---

[329] Cf. §3 para. 4 Nos. 1 and 4 *Chemikalienverbotsverordnung*: "...childproof fastenings conforming to the requirements of ISO 8317 (issued 1 July 1989)..." and "experimental lab kits...conforming to DIN EN 71 part 4, issued November 1990..." [translation by the author].

[330] Cf. also Falke 2000: 250ff. with further references.

[331] Cf. Denninger 1990: marginal ns 83ff., 132–3; Falke 2000: 9ff. for an overview of DIN procedures.

[332] *Vertrag vom 5. 6. 1975 zwischen der Bundesrepublik Deutschland, vertreten durch den Bundesminister für Wirtschaft, und dem DIN, deutsche Institut für Normung e.V., vertreten durch dessen Präsidenten.*

agreement obliges the DIN to have regard for the public interest when developing standards, place state officials on its committees, observe procedural provisions, give priority to standardization projects initiated by the government, and make its standards publicly available.[333] The technical standards issued by the DIN are not legally binding, since they originate with private actors. Nevertheless, such private technical standards become legally relevant, especially those concretizing indefinite legal conceptions. The administration can, for example, refer to private standards when determining the "best available technique" (Lübbe-Wolf 1991: 226).

A number of constitutional problems, including the issue of legitimacy, arise with the implementation of private technical standards in public law. First, the constitutional *Wesentlichkeitsgebot* again comes into play, with the consequence that a substantial part of any law must have parliamentary approval.[334] Standardization organizations are not legitimized through democratic elections (Mengel 2004: 147). Consequently, private standards lie outside the constitutional legitimacy framework (Lübbe-Wolf 1991: 233).

The setting of standards by private bodies is problematic for several reasons: economic interests are represented disproportionately compared to other interests like health or the environment; public participation in the development process is not always guaranteed; the committees that set such standards function as "closed clubs"[335] and their operations often lack transparency (Kloepfer 2003: 143–4).

It can be argued that, formally, democratically legitimized organs can ultimately decide on the validity and applicability of technical standards (Brohm 1987: marginal n.37); but this does not prevent a practical erosion of the constitutionally mandated level of legitimacy. As a consequence, other ways must be sought to compensate for this lack of democratic legitimacy in technical standards (Lübbe-Wolf 1991: 234–42).

One possible way of remedying this deficiency of to use an output-oriented approach. According to this perspective, technical standards are acceptable, because they reflect a consensus of the scientific community. Since these experts possess greater knowledge, they are better equipped to determine the proper course of action. In this process,

---

[333] For an overview of the agreement cf. Falke 2000: 61ff.

[334] Cf. also Federal Constitutional Court, Order of 9 May 1972: 158.

[335] Thus, a criticism from the US perspective points out that the German model overemphasizes engineering solutions at the cost of public participation, cf. Rose-Ackerman 1994: 1292.

they would be accountable to the scientific community alone.[336] This approach resembles that advocated by Schelsky. The same objections would apply in this case, namely, that the scientific process is ambiguous and uncertain. Scientists can err or neglect certain relevant societal values (Majone 1984: 15ff.). Placing decisions of health and safety into the hands of a technocratic elite is highly problematic and would likely be unacceptable to many.

Instead of focusing on an outcome, technical standards can be legitimized through the observance of procedural requirements: process, and not outcome, would determine their legitimacy (Majone 1982: 307; 1998: 20; Freiherr von Lersner 1990: 197). The question here is of how procedures can contribute to the legitimacy of the standard. In this respect, it is helpful to study the procedural elements of parliamentary lawmaking and their legitimizing effects.

Parliamentary laws have the highest legitimacy, since Parliament directly represents citizens and such laws are the outcome of a deliberative process between competing factions. Through debate the various arguments for and against a proposed measure are brought forth, discussed, and weighed. The whole process is public, which has the effect that the different arguments can be allocated to a specific group or person. Thus, it is clear to the public who is accountable for the final decision.

Elements of these deliberation processes also appear in administrative procedures. During the authorization procedure, for example, before the construction of a nuclear power plant, the various elements of the intended measure must be discussed.[337] In this process, citizens can bring forward their concerns, which must be considered by the authorizing agency. This process is necessary to ensure citizens' basic rights are protected.[338]

It is possible for a state to incorporate private technical standards into national law, if they satisfy the minimum standard of democratic procedures (*Sachverständigenrat für Umweltfragen* 1996: para. 897). If the minimum standard is not observed, technical standards cannot be included, giving the state influence over private standard setting to a certain extent (Mengel 2004: 148).

---

[336] From an American perspective this is a flaw of the German system, cf. Rose-Ackerman 1994: 1296.

[337] Art. 15 para. 1 of Council Directive 96/61/EC explicitly obliges the EU member states to take measures ensuring public participation prior to the authorization of an installation.

[338] Federal Constitutional Court, Order of 20 December 1979: 64–5.

The following procedural requirements can be distilled from court rulings and writings of legal scholars:[339]

- Technical standards specifying indefinite legal conceptions must comply with other legal provisions.
- Experts must give reasons for the technical standards.
- Experts from the scientific disciplines concerned must be involved in the standard-setting procedure.
- Standard-setting committees must have a mixed composition, allowing for the representation of the various affected interests.
- Standard-setting procedures must be transparent.
- Technical standards must be publicly available.
- Technical standards must be revised and regularly updated.

Regarding the first item on this list, it is clear that indefinite legal conceptions cannot be specified by illegal technical standards. In order to comply with the law, technical standards must observe the purposes of the provision they specify and the intended objects and levels of protection (Gusy 1994: 208). The latter aspect is of special importance. The level of protection is secured through the constitutional guarantee of certain basic rights and by laying down state purposes. For example, Art. 1 para. 3 of the *Grundgesetz* provides that all state organs must observe the basic rights guaranteed by the constitution. These rights are the reference point for measuring technical standards, since the administration and other state organs are duty bound to apply them.

The requirement that reasons be given for certain decisions is a common practice in administrative law.[340] There are some limitations to this requirement, since a certain level of flexibility is necessary. Moreover, it cannot be understood as an obligation to produce detailed documentation. Instead, the statement of reasons must include information on the empirical data used in the decision-making process. If the decision is based on valuations or uncertainties, these must be disclosed (Freiherr von Lersner 1990: 195–6). This obligation ultimately enhances the transparency of decision-making processes, and allows others to understand better the rationale of the decision (Majone 1998: 21), opening up

---

[339] Cf. the list of requirements in Mengel 2004: 148–9; Gusy 1994: 208ff.

[340] Cf. §39 para. 1 of the *Verwaltungsverfahrensgesetz*: "An administrative act given in written or electronic form…has to be issued with explanatory statements. The explanatory statement must include the substantial factual and legal reasons relevant to the agencies' decision.…" [translation by the author].

the matter for review and debate. Committees have to anticipate such public scrutiny and are encouraged to have solid reasons for their decisions, which is why the requirement to give reasons has not only an *ex post* function, but also has an *ex ante* component.

The standard-setting procedure must also involve experts from all disciplines affected by the standard.[341] This means that committee members should be required to produce proof of their expertise and disclose relevant personal, legal, or economic relationships (Freiherr von Lersner 1990: 196). For example, university professors serving on these committees must reveal information on the origins of third-party funds. This helps to identify possible conflicts of interest.

A standardizing committee made up of a diverse membership allows for the exchange of competing ideas.[342] The critique raised against the Federal Constitutional Court's concept of legitimacy becomes relevant here. From a constitutional perspective, when the legitimacy chain weakens so much that the executive's decision-making process lacks parliamentary backing and is as a resultlegitimacy deficit, civil society participation is not only permissible, it is mandatory (Fisahn 2002: 247; Schuppert 1977: 399). The participation of many different stakeholders guarantees that outside perspectives will be advanced and taken into account, and provides additional material for open debate (Lübbe-Wolf 1991: 243; Denninger 1990: marginal ns 178–9). However, the composition of a committee must be balanced, with individuals representing many critical viewpoints.[343]

The Federal Constitutional Court has stated that one prerequisite of democratic legitimacy is that decision-making procedures of state organs must be transparent. Citizens must be able to see and understand the processes in order to be able to communicate, get involved, and not be condemned to passivity until election day.[344] In order to make standard-setting processes traceable for outsiders and determine the accountability of actors for their activities, the committees and procedures must be designed transparently. This is achieved by publicly announcing sessions early enough for interested parties to consider their participation

---

[341] Cf. Federal Administrative Court, Judgement of 29 April 1988: 264.

[342] Cf. §§48 and 51 of the *Bundes-Immissionsschutzgesetz*. Prior to the enactment of certain administrative regulations, the parties concerned must be heard. This includes the scientific community and business associations. Cf. also Denninger 1990: marginal n. 59; Führ 1993: 100.

[343] A negative example in this regard was the composition of *AGS* and *KTA* in the 1980s, cf. Winter 1986: 16.

[344] Federal Constitutional Court, Judgement of 12 October 1993: 185.

and elaborate their standpoint and by publishing the agenda, drafts, and statements, thereby allowing public review (Schulte 1998: 179).

Requiring that standards be made available to the public ties in with the need for transparency. Although likely not always as fundamental as the guarantee established in Art. 1 para. 2 of the EU Treaty, which obliges organs of the EC to act transparently, every democratic constitution is built on the principle of transparency. As Parliament acts as the ultimate supervisor, parliamentary oversight of government power must be traceable and unconcealed from the public (Bröhmer 2004: 15). For example, according to Art. 42 para. 1 of the *Grundgesetz*, debates at the *Bundestag* have to be public. Certain basic rights are intrinsically linked to the principle of transparency; for example, Art. 5 para. 1 of the *Grundegesetz*, which guarantees the freedom of speech, has an informational component, regulating unrestricted access to information (Id.: 196ff., 206ff.). The public availability of session documents and draft standards make the standard-setting process more transparent. Transparency can be achieved by regularly publishing the relevant documents, either in print or on the Internet.

It is self-evident that technicals standards have to be up to date and take into account the latest scientific findings (Gusy 1994: 209; Majone 1984: 19). After all, this is one of the reasons why the state's legal system incorporates private technical standards. Technical standards can be kept up to date through periodical review. In addition, the technical standards must be open to revisions, which means that procedures must be in place that allow for objections and revisions to the standard.

In Germany, a general law on the setting of technical standards has not been enacted yet; instead, elements are scattered throughout the legislation. Noteworthy in this regard is the failed attempt to create an Environmental Code for Germany, through the merging of the essential elements of environmental law (Ministry for the Environment, Nature Conservation, and Nuclear Safety 1998). The draft (*Umweltgesetzbuch – Entwurf der unabhängigen Sachverständigungkomission, UGB-KomE*) contained a section dedicated to "Rule Making and Regulation" (§§11–40 of *UGB-KomE*) and §§25–30 *UGB-KomE* laid out detailed rules for administrative rulemaking. The Federal Government would be required to state reasons (§§29, 16 para. 1 of *UGB-KomE*) and hear representatives of the Environmental Committee, a specific body comprised of representatives from the scientific community, industry, associations, state and local authorities, and other societal groups (§§29, 18 para. 1 of *UGB-KomE*), and have the right to make proposals (§§29, 19 of *UGB-KomE*).

Under this law, all administrative rules would be subject to periodic review (§30 *UGB-KomE*).

§§31ff. of *UGB-KomE* laid down requirements for the reference to technical rule issued by civil society actors. Significant in this regard is §32 of *UGB-KomE*, which sets out the procedure for implementing technical rules. Technical rules have to fulfil certain prerequisites to be introduced into legislation, including: being compliant with environmental provisions, produced with input from expert agencies, technically sound and balanced, accompanied by adequate reasons, and adopted through an open and public procedure with participation of representatives of the competent highest federal authorities. Technical rules would will to be published and made accessible. Most importantly, their incorporation into the legal order must be within the public interest.

A proposed law on technical standards, prepared in 2003 by an ad hoc "Commission on Risks," has also not been realized. This proposal was the outcome of discussions on restructuring and improving risk assessment and management in Germany, and included many of therequirements for the creation of technical standards mentioned above (Risikokommission 2003: 71ff.). Building on the work of the *UGB-KomE*, the Commission on Risks suggested that certain basic rules be instituted with regard to setting technical standards (§4 of the proposal). Standards must be set in order to prevent or minimize health and environmental risks and with consideration to the type, extent, probability, and reliability of the predictions (*"Aussagesicherheit"*) of the specific risks. A safety margin between standard and expected harm must also be included.

In order to ensure the transparency of the whole procedure, the responsible agency is required to publish the draft regulation that implements the standard (§8 of the proposal). This opens up the process to further comments on the draft, and allows people to voice their objections (which the agency must take into account). If it intends to depart from these objections, it is obliged to provide reasons for doing so. Furthermore, the standards themselves must contain provisions on their revision (§9 of the proposal).

In US administrative law, §553 Administrative Procedures Act (APA) is the central provision on administrative rulemaking. It is more general than the proposed rules on standard setting in Germany. The provision outlines a process for administrative rulemaking that ensures transparency and public participation, contributing to the legitimacy of the rules.

"Rules" are defined in §551 (4) of the APA as "...agency statements of general or particular applicability and future effect designed to implement, interpret, or prescribe law or policy...." According to §553 (b) of the APA, an agency must, except in specified cases, publish a general notice of proposed rulemaking in the Federal Register, stating the time, place, and proceedings, give reference to the legal authority under which the rule is proposed, and state the terms or substance of the proposed rule or describe the subjects or issues involved. The notice is meant to inform the public and elicit comments. After publishing the notice, the agency must give interested persons an opportunity to participate in the rulemaking process (§553 (c) (1) of the APA). Furthermore, under §553 (c) s. 2 APA, the agency must incorporate a concise general statement of the basis and purpose of the rules into the regulation (i.e. state reasons for their adoption). These general provisions on the rule-making process are complemented by special laws (for example, §4 (b) (5) of the TSCA), which require that interested persons must be given an opportunity to present their views, arguments, and any additional information orally or in writing. Furthermore, under §4 (b) (2) (B) of the TSCA, the Administrator must regularly review the adequacy of testing rules, and, if necessary, initiate a revision process. Eventually, the rule is published in the Code of Federal Regulations (CFR), making it accessible to the public.

Technical standards in supranational settings

A survey of the comitology system in the European context provides a different picture on the setting of technical standards. The fact that the system takes place in a supranational context adds an extra level of complexity. The committees are made up of delegates from member states, whose role involves the development and review of EC legislation. These committees are an example of close cross-border administrative cooperation, and bear some semblance to the transnational bureaucracy networks.

The practice of comitology raises legitimacy concerns, since a large part of the legislative process is placed in the hands of "Eurocrats." As a result, it is worth examining the problem and the proposed remedies, in order to gain insights on how to address the legitimacy problem of transnational public law.

It is first necessary to explain the background and practice of comitology. While it is clear that comitology has to do with committees, it does not refer to all committees that exist at the community level. The "grand" committees envisaged in the EC Treaty, that is, the Economic and Social Committee, the Committee of the Regions, and the Committee of Permanent Representatives do not fall within the scope of comitology, nor do expert groups set up by the Commission.[345] Rather, the term applies to the various committees established on the basis of Art. 202 third indent of the EC Treaty.[346] According to this provision, the Council confers on the Commission the necessary powers to implement the rules set out in the Treaty. Since the Council can impose requirements on the exercise of these powers, it may assign committees of representatives from the member states to the Commission. Comitology thus applies to committees that are established through secondary Community law to assist the Commission in implementing Community legislation.

Committees comprised of national officials have been established at the European level since the early 1960s.[347] Almost as soon as the Council delegated implementing powers to the Commission, committees of national civil servants were established to assist with the implementation process. The first of these so-called management committees were established in the area of agriculture policy (Bradley 1992: 694). Regulatory committees, which were empowered to implement permanent rules, followed a couple of years later, in the late 1960s (Bergström 2005: 81ff.). Today, there are almost 250 committees, covering every Community policy area.[348] The majority of committees – especially in the policy field "Environment" – operate under the regulatory procedure.[349]

The function and purpose of comitology has changed over the years. In the beginning, through the establishment of committees, member states aimed to control the Commission's operations (Hartley 2003: 24; Hummer 1998: 285). However, committees also gave the Commission the opportunity to develop ties to national

---

[345] Report from the Commission on the working of committees during 2004 (COM(2005) 554 Final): 7.

[346] ECJ, Judgement of the Court of First Instance, 19 July 1999, para. 57; Cf. also Bergström 2005: 2.

[347] For an overview of the development of committees through the last four decades cf. Hummer 1998: 287ff.; Vos 1997: 21–2.

[348] Cf. Report from the Commission on the working of committees during 2004 (COM(2005) 554 Final): 8; cf. also European Commission, List of Committees which Assist the Commission in the Exercise of Its Implementing Powers. Before the Commission issued reports on committees, it was apparently very difficult to assess an exact number. Hence, Töller 2002: 22 speaks of roughly 400 committees.

[349] Report from the Commission on the working of committees during 2004 (COM(2005) 554 Final): 10.

administrations (Bergström 2005: 54). The relationship between the Commission and the committees changed in the wake of a ruling handed down by the European Court of First Instance. The Court concluded that the committees were not independent of the Commission. Comitology committees do not have their own address, administration, budget, or archives.[350] Instead, their operations are closely connected to the Commission's work. The Commission provides secretarial services, prepares meetings, and determines the agenda (Falke 1996: 128–9). Consequently, in regards to the issue of the case, which concerned the access to committee documents, the Court held that committees fall under the jurisdiction of the Commission.[351] This ruling transformed committees from "control bodies to structures subordinate to the very institution they were supposed to control" (Dehousse 2003: 808).

Apart from these institutional considerations, there are a number of reasons for the Council to leave the concretization of Community legislation to the Commission and committees. In fact, these reasons do not differ much from the reasons why legislators delegate the concretization of laws to the administration at the national level. First, like the national legislators, the Council does not always command the necessary expertise and consequently leaves the technical aspects of the regulatory measures to experts (Dehousse 2003: 799–800). This allows the technical aspects of the legislation to be kept up to date. The reasons for delegation do not differ much from the national legislator's motives for delegating technical matters to the executive or third parties (for example, standardization organizations). Political considerations might also inspire the Council to leave contentious matters, which cannot be settled easily at the highest political level, to committees (Dehousse 2003: 799; Roller 2003: 251). In addition, the Commission's limited resources force it to rely on outside expertise. Civil servants can easily provide this expertise, contributing a practitioner's perspective as the parties actually implementing EC law (Winter 1996: 109; Dehousse 2003: 800).

The matter of delegation of powers was subject to judicial scrutiny as early as 1970. The ECJ found "that ... it cannot be a requirement that all the details of the regulations ... be drawn up by the Council ... ." It is sufficient that the Council defines the basic elements of the matter to be

---

350 ECJ, Judgement of the Court of First Instance, 19 July 1999: 2483, para. 58.
351 ECJ, Judgement of the Court of First Instance, 19 July 1999: 2484, para. 62.

concretized in the authorizing act.[352] This falls below the requirements for delegation of legislative powers under Germany's constitutional law. It is neither comparable to Art. 80 para. 1 of the *Grundgesetz*, nor is it analogous to the *Wesentlichkeitsgebot*, under which Parliament must regulate substantial matters relating to basic rights. Instead, the ECJ considers only the key aspects of Community policy as substantive elements.[353]

Details of comitology are regulated in Council Decision 1999/468/ EC,[354] which is complemented by Standard Rules of Procedures issued by the Council[355] and an Agreement Between the European Parliament and the Commission,[356] which form the basis of the current Comitology system. The current system is the result of a long and complicated row between the Parliament and the Council (Bergström 2005).

The two most interesting procedures for the purpose of this investigation are those provided for in Arts 5 (regulatory procedure) and 5a (regulatory procedure with scrutiny) of Council Decision 1999/468/ EC. The significance of these procedures stems from their impact on Community law and implementation by member states. Committees acting under these procedures actually set tertiary law, complementing the Communities secondary law (Winter 1996: 107). Often, it is their task to adapt Community legislation to the technical or scientific progress, that is, review the technical provisions and change them. As a result of the technical character of these provisions and the scientific expertise involved, these regulatory procedures are comparable to the setting of technical standards by transnational bureaucracy networks.

Under the regulatory procedure, the Commission prepares a draft measure and submits it to the committee for approval. Within a certain time limit the committee must submit its opinion, which must be delivered by the majority according to Art. 205 paras 2 and 4 of the EC Treaty. If the committee concurs, the Commission adopts the measure

---

[352] ECJ, Judgement of the Court 17 December 1970: 1170, para. 6; for a thorough discussion of the ruling cf. Bergström 2005: 97–104.

[353] ECJ, Judgement of the Court 27 October 1992: 5434, para. 37; Cf. also Roller 2003: 260.

[354] Council Decision 1999/468/EC. Cf. Bergström 2005: 249–84 for an in-depth analysis of the decision's evolutionary history.

[355] Standard Rules of Procedure – Council Decision 1999/468/EC.

[356] Agreement Between the European Parliament and the Commission on procedures for implementing Council Decision 1999/468/EC of 28 June 1999 laying down the procedures for the exercise of implementing powers conferred on the Commission.

as envisaged. If the Commission disagrees or fails to issue an opinion, the proposed measure is submitted to the Council. At the same time, the Commission informs the Parliament, which can consider the proposal. If Parliament then finds that the proposed measure exceeds the implementing powers set out in the basic instrument, it informs the Council of its position. Eventually, the Council must decide, within a specific time limit and with a qualified majority, whether to oppose or adopt the measure. If it opposes the measure, the Commission may resubmit the measure or propose other legislative measures. If the Council fails to act within the time limit, the Commission may adopt the measure.

In July 2006, the Council amended its Decision 1999/468/EC and introduced a so-called regulatory procedure with scrutiny.[357] This regulatory procedure differed from Art. 5 of Decision 1999/468/EC in that it required stronger parliamentary involvement. Once a proposal receives the committee approval, it is referred to the Council and Parliament for further scrutiny. If either body considers the proposed measures to exceed the Commission's implementing powers, or views them to be incompatible with the purpose and content of the basic legal act or a violation of the principles of subsidiarity or proportionality, it can oppose the draft.[358] The Commission may then amend its proposal and resubmit it to the committee.[359] If the committee rejects the Commission's proposed measure, the Commission must submit the measure to the Council and forwards it at the same time to the Parliament for scrutiny[360]. In this case, Council's opposition of the measure results in it not being adopted.[361] However, if the Council intends to adopt the measure, it informs the Parliament. Parliament may reject the measure on the grounds already mentioned – an excess of implementing powers, incompatibility with the aim or content of the basic legal act or violation of the principles of subsidiarity or proportionality. The Commission may then resubmit an amended proposal or propose alternative legislative measures. If the Parliament fails to act within the given time limit, the measure can then be adopted.[362]

The committees operating under these two regulatory procedures exercise a certain amount of authority. The activities of committees operating under the regulatory procedures make a large contribution

---

[357] Art. 1 No. 7 Council Decision 2006/512/EC.
[358] Art. 5a para. 3 (b) Council Decision 1999/468/EC.
[359] Art. 5a para. 3 (c) Council Decision 1999/468/EC.
[360] Art. 5a para. 4 (a) Council Decision 1999/468/EC.
[361] Art. 5a para. 4 (c) Council Decision 1999/468/EC.
[362] Art. 5a para. 4 (d)–(e) Council Decision 1999/468/EC.

to Community law by establishing tertiary law. The outcome of the comitology process eventually results in amendments to European legislation and ultimately, the compliance of European citizens with this law. Thus, the question is of why this authority is acceptable (Falke and Winter 1996: 569; Roller 2003: 259).

First, the matter of the legitimacy of comitology touches upon the institutional balance of the Community organs. It has been mentioned above, that Decision 1999/468/EC ended an intense conflict between the Parliament and the Council. The roots of this dispute lie in Art. 202 third indent and Art. 251 of the EC Treaty. In 1986, the Single European Act introduced the co-decision procedure, thereby upgrading the role of the European Parliament to "co-legislator." However, while it is involved in the creation of the basic legal act, it played only a minor role in the implementation of this act through concretizing measures adopted by committees (Hummer 1998: 286–7). Consequently, the Council's first "Comitology Decision"[363] led to "one of the most acute interinstitutional clashes in the Community's history..." by completely dismissing Parliament's demand for an acknowledgement of its role in the legislative process (Bradley 1992: 695; Hummer 1998: 303–4; Vos 1997: 218–19) and the Parliament fought fiercely for its involvement in the process of delegated lawmaking.[364]

The second comitology decision does a better job of integrating Parliament into the legislative process (Roller 2003: 272). According to the Council's comitology decision and the Parliament's Agreement with the Commission, the European Parliament is to receive agenda, drafts, voting results, and summary records of committee meetings from the Commission.[365] Nevertheless, it is important to note that Art. 8 of Council Decision 1999/468/EC gives Parliament the ability to intervene in the committee process. Moreover, under the recently introduced regulatory procedure in Art. 5 (a) of Council Decision 1999/468/EC, the role of the European Parliament is equal to that of the Council.

There was also a second conflict between the Commission on the one hand and the Council and Parliament on the other. In a case adjudicated by the ECJ in 2003, the Commission disapproved of a derogation from the procedures laid down in Council Decision 1999/468/EC

---

[363] Council Decision 1987/373/EEC.

[364] However, Parliament's effort to settle the conflict by referring it to the ECJ failed due to the application being found inadmissible, cf. Judgement of the Court, 27 September 1988: 5615.

[365] Cf. Art. 7 para. 4 of Council Decision 1999/468/EC and para. 1 of the Agreement.

in a basic legal act adopted by the Council and the Parliament, which would effectively weaken its own position. The Court pointed out that while Council Decision 1999/468/EC is not a legally binding instrument, it has legal effects in establishing a practice for institutions to follow. While these bodies may depart from this practice, they must give reasons for doing so.[366]

The struggle of the Community organs for an interinstitutional balance conveys an impression of control over the comitology process. However, Parliament and the Council only become involved in the process if a committee intends to reject a proposed measure. This is rarely the case: usually, committees agree to the proposed measures.[367] This means that the Council and the Parliament rarely have the opportunity to intervene. In addition, recent observations indicate that the committees have increasing autonomy (Dehousse 2003: 799; Roller 2003: 261–2). In view of the fact that their original function was to control the Commission, one can indeed quote Juvenal's famous question: *sed quis custodiet ipsos custodes* – but who watches the watchmen?

In view of the Commission's changed relation to the committees, which is also expressed in the mentioned above suit in reaction to the Council's and Parliament's deviation from Council Decision 1999/468/EC, it appears that the Commission and the committees jointly form the complex of comitology. Hence, the legitimacy of comitology becomes an issue. In particular, scholars have extensively criticised the lack of transparency of committee procedures as a delegitimizing factor (Roller 2003: 262).

In terms of formal legitimacy, a committee's existence is legitimized through its founding act, which is legitimized through its legislative origins and conformity with the EC Treaty. The legal act comes into being through the Council acting alone or jointly with the European Parliament, in accordance with Art. 251 of the EC Treaty. The Council consists of representatives from national governments, each legitimized by their national parliament. The members of the European Parliament are collectively legitimized through the elective acts of European citizens (Höreth 1998: 8). Committee members, representatives from the

---

[366] ECJ, Judgement of the Court, 21 January 2003: 998–9, paras. 50–1.

[367] For 2004, the Commission reported 17 referrals to the Council for regulatory and management procedures. This amounts to 0.5 per cent of all implementing measures under these procedures. Cf. Report from the Commission on the working of committees during 2004 (COM(2005) 554 Final): 6.

administrations of member states, link to their respective national legitimacy chain.

It is clear this formal approach misses the point, by failing to address the transparency issue so that once the Council and the European Parliament have delegated the powers, they are not required to control how these powers are exercised. One of the duties of the European Parliament is to supervise the Commission (Arnull et al. 2000: 36–7). As a consequence, the control of the committee practice would clearly fall within this role, especially since the work of such committees impinges on the Parliament's legislative role. This explains the Parliament's discontent with the first comitology decision and its struggle for greater opportunities for involvement in the legislative process.

The European Parliament may invoke Art. 1 of the EU Treaty, which explicitly states "…decisions are taken as openly as possible and as closely as possible to the citizen." In this way, the current law requires that the Commission inform Parliament and publish the agenda and minutes of committee meetings.[368] Although this enhances the transparency of comitology, the publication of committee documents is met with scepticism on the grounds that monitoring committee work is not feasible, and, particularly, that parliamentarians cannot process this flood of information (Dehousse 2003: 805–6). Nevertheless, it is important to note that a large part of the committee work is publicly available, allowing the public to track the work of the committees. It can potentially make known maladministration, resulting in greater parliamentary involvement.

The legitimacy problem posed by comitology can also be approached from the procedural side. All committees must draw up procedural rules based on the standard rules of procedure issued by the Council. These ensure that the committee proceedings follow certain protocols, making them more predictable. The rules allow for the admission of third parties, that is, experts or representatives of other states or organizations, to committee meetings. Third parties cannot be present during voting.[369] The committees also publish committee documents, but the actual discussions remain confidential.[370] Although this is a starting, letting civil society actors participate in committee meetings alone is not sufficient to convey a significant level of legitimacy. While, in

---

[368] Art. 7 para. 2 of Council Decision 1999/468/EC; for this purpose, the Commission maintains a "Register of Comitology" on the Internet, cf. European Commission 2008.

[369] Cf. Art. 8 of the Standard Rules of Procedure.

[370] Cf. Art. 14 of the Standard Rules of Procedure.

principle, this ensures that concerned parties can be heard, in practice, weaker interests (for example, environmental groups) typically lack the funds and human resources to participate in every meeting. Thus, one danger of this type of participation is that stronger interest will be over-represented (Falke and Winter 1996: 571).

In addition to improving this aspect of public participation, other elements can also contribute to the legitimacy of comitology. One such aspect is the scientific expertise concentrated in the committees (Id.: 569–70). Although many misjudgements in the past 40 years have tarnished scientific expertise as a legitimizing factor, most people still view a consensual vote of experts as acceptable. Another procedural feature that adds to the legitimacy of comitology is the clash and resolution of different risk cultures. Committees bring together officials from all member states, which have their own unique perceptions of the risks and views on how to resolve specific problems. This guarantees that a host of different ideas are heard when seeking a practical solution to a problem (Id.: 571). In the end, EC Treaty provisions ensure that committees do not settle for the lowest common denominator when seeking solutions and elaborating standards, but aim towards a high level of protection.[371]

In sum, the mechanisms for ensuring legitimacy of regulatory measures taken in the EU are similar to those employed in Germany in regard to technical standards. The main difference here is that there is a strong institutional structure, reflecting great interest in the work of comitology committees, as demonstrated by Parliament's struggle for influence. Apart from that, transparency, public participation, scientific expertise, and observation of values are elements that can collectively contribute to the legitimacy of highly technical decision-making processes.

Conclusion

There are proven and tested strategies to legitimize the setting of technical standards in national and supranational environments. Input legitimacy, conveyed via Parliament, is insufficient and needs to be supplemented by other strategies, including:

---

[371] Cf. Arts 2, 3 para. 1 (c), 61(e), 95 para. 3, 152, 153, and 174 para. 2 of the ECTreaty, all of which oblige the Community to pursue high levels of protections in its consumer, environment and health policy and high levels of security in police and judicial matters, respectively.

- accordance with legal provisions that lay down the purpose of the technical standards and the level of protection that the standards must aim to preserve;
- the observation of values;
- the requirement to give reasons for decisions, outlining the decision rationale, the empirical data used in the decision-making process, and the disclosure of valuations and uncertainties;
- the decision-making body must be composed of experts, guaranteeing that the outcome of the decision-making process is based on expertise;
- the committee must have a diverse composition, allowing representatives of different risk cultures and scientific backgrounds to share ideas and find common solutions;
- the committees must allow those affected by the decisions and those representing relevant interests to voice their opinion;
- the workings of the committee must be transparent and predictable;
- the decisions must be made publicly available; and
- the decisions must be revisable and regularly reviewed regarding their currency.

The role of the European Parliament in the comitology procedure shows that this process does not result in Parliament being redundant. In fact, there are ways for parliamentary involvement in the setting of technical standards. Even though parliamentarians might not be able to scrutinize every procedure effectively, they would have the opportunity to intervene in cases of gross maladministration. Provided there is adequate transparency and reporting of committee activities, civil society representatives can play a major role, tracking the decision-making processes, pointing out errors, and calling for legislative intervention in significant matters.

17
Legitimacy of Transnational Public Governance

After establishing a set of criteria to determine the legitimacy of technical decisions and decisions in environments beyond the state, these measures can be applied in investigating issues of legitimacy in transnational public governance. However, one must first examine the extent to which legitimacy is necessary to transnational public governance.

Networks have certain shortcomings that affect their legitimacy. Because of their informal character and they way they operate at the transnational level, networks have been diagnosed with a "chronic lack of legitimacy."[372] They run the danger of becoming dissociated from the individuals who are affected by their decisions – a global technocracy operating under the radar of national parliaments in spite of the significant impact of their decisions.

Transnational public law, while not being obligatory in the positivist sense, is based on authority. This authority has to be legitimized. The question then is of how transnational public law on chemical safety is legitimized.

The decisions taken by transnational bureaucracy networks pass through several levels and are ultimately set by formal decision-making organs belonging to the IO that created the network node. However, it would be wrong to begin the legitimacy discussion here; since IO organs adopt the decisions taken by transnational bureaucracy networks in an unaltered form, they must be legitimized from the outset.

In the field investigated here, the matter of legitimacy is an important issue in the following areas.

---

[372] *Supra* 42.

The first are the IPCS and its CICADs. Although CICADs are not transnational public law, they do have legal relevance. They are based on a consensus of international experts, giving them special weight. National risk management measures tie in with CICADs but because the latter are not legally binding, national risk management agencies can deviate from their content although when doing so, they must go against international expert consensus, which might be extremely difficult. Moreover, since the IPCS is based on a tripartite agreement and is connected to numerous national agencies, it has a unique institutional character and is paid significant attention.

The other three are the standards that have already been discussed in this book: the Test Guidelines, GHS and UNRTDG. As transnational public law, they are incorporated into national legal orders and thus influence individual states' approaches to chemical safety and in the end have an impact on the activities of private actors.

The input legitimacy of the IPCS is relatively weak for several reasons. First, due to its unique structure, the IPCS has become quite independent from its parent organizations (the WHO, ILO, and UNEP). Second, from the perspective of the German Parliament as an original source of legitimacy, the IPCS' input legitimacy is faulty. The WHO has been the dominant force behind the establishment of the IPCS and is highly influential. Germany acceded to the WHO under an administrative agreement that did not require parliamentary consent, because it was believed that it would not affect Germany's political relations.[373] However, in the IPCS a body exists which was created by and is sustained with major contributions from the WHO, whose decisions impact Germany's chemicals policy and risk management assessments. As a consequence, today WHO activities may very well affect Germany's political relations. Another very good example are the powers of the WHO that stem from International Health Regulations. During the 2003 outbreak of the Severe Acute Respiration Syndrome, the WHO demonstrated its powers when it advised against travel to Toronto (Fidler and Gostin 2006).

One could argue that the *Bundestag* should be aware of the IPCS and its activities, because it sanctioned Germany's financial contributions to the organization in its annual budget. However, it is unclear whether Parliament actually discussed each and every item in the budget.

---

[373] Cf. *Supra* 196 on administrative agreements.

Furthermore, the activities of the IPCS and CICADs, in particular, have never been discussed in parliamentary committees so it is doubtful that Parliament is aware of the IPCS and its activities.

The legitimacy chain is another argument, which could be invoked to support the existence of some type of input legitimacy. However, the objection that the legitimacy chain is weakening with respect to European or international activities is also relevant in this transnational context. This is also the case in transnational environments, especially since attached officials operate in increasingly autonomous structures, operating in an environment of which Parliament does not seem to be fully aware. As a consequence, Parliament alone cannot be considered to be a sufficient legitimizing source of the IPCS and, furthermore, input legitimacy is ineffective in this context.

The IPCS must now be examined with regard to the application of the alternative legitimacy strategies described above. There was no indication in the documents reviewed that the IPCS and its activities violated any provisions of international law. But to what extent does the observation of values play a role? The MoU founding the IPCS merely outlines the mode of the tripartite cooperation, and does not impose a duty on it to observe certain objectives or values. The Guidelines for the Preparation of CICADs do not include such requirements either. However the values guiding the IPCS' activities are laid down in the founding documents of its parent organization. Art. 1 of the WHO Constitution sets out a key objective that there "...shall be the attainment by all peoples of the highest possible level of health." Similarly, in the preamble of the ILO Constitution, the ILO has a mandate to strive for social justice and humane labour conditions. Among UNEP's objectives, laid down in Art. 1 of the UNGA Res. 2997 (XXVII), is the promotion of international cooperation in the field of environmental protection.

Eventually, these objectives led the WHO, ILO, and UNEP to create the IPCS. In fact, these must also be the valid guiding principles for its work. Nothing in the empirical data gathered in the course of this study suggests that CICADs or any other IPCS activity did not adhere to these objectives.

Another legitimizing element is the practice of giving reasons for decisions. There are two important things to note regarding CICADs. First, the procedure for the elaboration of CICADs follows strict scientific standards. A document is peer reviewed, and later examined by the Final Review Board, so that it is thoroughly evaluated before being published. The author has to justify the content of the draft to the peer reviewers and the Final Review Board. Second, if the Final Review Board cannot achieve a consensus, the dissenting opinion will be published in the CICAD.

The scientific value of a CICAD eventually depends on the composition of the bodies involved in the elaboration process. CICADs are held in high regard because they represent an expert consensus. While national authorities are permitted to send experts to the deliberations, they must submit a "Declaration of Interests" to ensure their neutrality. This guarantees that the CICAD is actually based on expertise. Those affected by the CICAD – particularly those companies manufacturing the substance – are also given a voice in the drafting process. Excessive lobbyism is avoided by restricting the participation of civil society actors to the Final Review Board meetings.

As regards the transparency and predictability of the decision-making process, the IPCS operates on the basis of a clear set of procedures laid down in the MoU establishing the organization. This is also the case for the development process of CICADs. This process follows a set of guidelines issued by the IPCS. The IPCS Risk Assessment Steering Group, which plays a significant role here, has also received ToR guiding its work. This enhances the predictability of the overall process. In addition, CICADs are made public on the Internet. The protocols governing the deliberations of the Final Review Board are also available. Thus, the public has access to all of the relevant documents and can get an idea of how the process functions. As a consequence, the IPCS is quite transparent, especially in comparison with IFROs.

A CICAD records what is known about a particular substance at a specific point in time. This means that, over time, as scientific knowledge expands, they can become obsolete. However, it is not really necessary to furnish CICADs with a "sunset clause" or call for an obligatory review. CICADs include only a basic set of information, requiring the user in any case to do additional research. CICADs serve as the starting point for risk assessment or risk management – they do not contain the actual measure. Hence, although the information presented in a CICAD must be factual, it does not have to be current.

To summarize, the IPCS and CICADs are examples of successfully applied alternative legitimacy strategies. They are widely accepted in the scientific community and among risk regulators on the basis of their factual content.

From a German perspective, the degree of input legitimacy for OECD decisions is more acceptable, because Germany's membership in this organization is based on a legislative act. In addition, Germany's contributions to the OECD Chemicals Programme are registered in the annual budget; and the Federal Government's report on animal protection, which was submitted to the Parliament, briefly mentions the participation of German officials in the creation of these standards.[374] However, it is doubtful that this is sufficient, since, in both cases, Parliament did not concern itself with details of the OECD Chemicals Programme. Furthermore, the argument of the weaking legitimacy chain would also apply in this case.

In view of the absence of a proper legitimization by Parliament through an unbroken and strong legitimacy chain, it is again necessary to discuss the existence of other legitimizing elements. The general aims and purposes of the WNT, which elaborates the Test Guidelines following a detailed procedure laid down in a guidance document, follows from Art. 1 of the OECD Convention. According to that provision, all OECD activities should aim to achieve the highest possible levels of sustainable economic growth and employment, as well as raise the standard of living in the member states, contributing to sound economic growth and the expansion of world trade. References to terms such as "sound" and "sustainable" indicate that OECD does not merely serve economic aims, but that it also takes account the social and environmental aspects of economic growth. This also expressed in Recital no. 8 C (81) 30, which recognizes the need for concerted action by OECD members to protect humanity and the environment from hazardous chemicals. These values also guide the work of the WNT.

The process in which the Test Guidelines are created is highly contentious, bringing experts from OECD member states together at the WNT. Drafts undergo an extensive review process, so that, at the end of the process, plenty of material exists documenting why the particular decision was made, which valuations were necessary, and which uncertainties needed to be dispelled.

---

[374] *Deutscher Bundestag, Unterrichtung durch die Bundesregierung: Tierschutzbericht 2003 – Bericht über den Stand der Entwicklung des Tierschutzes (BT-Ds. 15/0723): 74; Unterrichtung durch die Bundesregierung: Tierschutzbericht 2005 (BT-Ds. 15/5405): 35ff.; 53ff.*

The WNT is made up of experts from OECD member countries. This means that no special care is placed on the issue of admitting experts holding opposing views. However, it has been pointed out in the case of EC committees, that representatives from different countries have different backgrounds coming from different risk cultures. This ensures that a variety of views are represented in the WNT. The differences among risk cultures eventually guarantee a pluralism that can serve as a legitimizing element.

Civil society actors – particularly those with a scientific background – have the opportunity to review drafts, and sometimes OECD Secretariat requests their opinion. In addition, the procedures envision that they can – via the National Coordinator – initiate the process to develop or update a Test Guideline. Moreover, with the publication of the final product (i.e. Test Guidelines) – through OECD (they become part of C (81) 30), the proceduralization of the elaboration of Test Guidelines and the participation of societal actors, the whole process is transparent. Interested parties can watch, discuss and participate in the process. Public scrutiny of the work of the WNT is thus guaranteed.

In order ensure the Test Guidelines are up to date, the Guidance Document sets out a procedure to review of Test Guidelines, which might result their amendment or their removal. Therefore, they satisfy the requirement of revisibility and currency.

As far as the input legitimacy of GHS is concerned, one has to distinguish its two phases: first, its initial creation under the auspices of CG/HCCS and, second, the continued work on GHS by the UNSCEGHS. The *Bundestag* took notice of GHS only as a result of the government's reports on animal welfare, which were only briefly mentioned in the context of alternative methods.[375] The reference of GHS in a document dealing with a matter that only marginally touches upon the issue of chemical safety suggests that the Parliament did not yet really have the opportunity to understand the impact of GHS. It also failed to take notice of the elaboration process, and did not deal with the continued revisions under the UNSCEGHS. Once again, the legitimacy chain linking the German officials participating in the network structures with the Parliament shows strain. As a consequence, GHS and

---

[375] *Deutscher Bundestag, Unterrichtung durch die Bundesregierung: Tierschutzbericht 2001 – Bericht über den Stand der Entwicklung des Tierschutzes (BT-Ds.14/5712): 70; Unterrichtung durch die Bundesregierung: Tierschutzbericht 2003 – Bericht über den Stand der Entwicklung des Tierschutzes (BT-Ds. 15/0723): 74.*

the work of the CG/HCCS and UNSCEGHS must be legitimized by other means.

In that the consideration of values is a required element of legitimacy, the set of principles devised by the CG/HCCS in guiding the work on GHS becomes relevant. One goal was that the harmonization of classification and labelling standards would not lead to a reduction of the level of protection. In addition to this clear "mission statement," the objectives as laid down in the constitutions of the participating organizations also apply. The UNSCEGHS was set up by ECOSOC, and thus, operates within the UN framework. The principles guiding the working procedures within the UN as laid down in Art. 1 of the UN Charter also apply to the work of the UNSCEGHS.

Elaborating and maintaining GHS has been and continues to be an arduous process, dominated by the exchange of scientific views. The actual document is very detailed, and clearly sets out the rationale for its decisions. Unfortunately, the consultations with the CG/HCCS are not publicly available.[376] This obstacle notwithstanding, the work on GHS has been documented in specialized journals, and key documents, including GHS and UNSCEGHS agendas, are available on the Internet. As a result, the unavailability of certain documents does not really hamper publication and the overall transparency of the work on it.

In the formation phase of GHS, government officials were dispatched from participating countries and organizations to the various working groups. The documents from this process do not indicate whether the admission of experts was based on creating working groups of a diverse composition. Presumably, the fact that the experts came from several different risk cultures meant that opposing views were voiced and discussed. ECOSOC determines which countries (and, thus, which experts) are represented in the UNSCEGHS. In principle, ECOSOC could guarantee committee diversity. However, once again, the diversity of viewpoints will most likely already follow from its multinational composition.

NGOs have participated in GHS from its onset. For example, representatives from the ICEM and WWF sat on the CG/HCCS. The participation of civil society actors in the UNSCEGHS follows ECOSOC's RoP. Consequently, NGOs have a consultative status and can participate in discussions.

---

[376] Cf. ILO 2008b. Access to the relevant documents requires a password.

GHS was envisioned as a dynamic instrument. The UNSCEGHS was specially established to provide for the revision of GHS, thereby ensuring its revisibility and topicality.

Although the UNRTDG have existed for several decades and have an enormous impact on national and international law, the *Bundestag* has never taken notice of them. In addition to the weak legitimacy chain, the input legitimacy of the UNRTDG is rather poor.

Because of the institutional and topical closeness to GHS, the applied alternative legitimacy strategies are quite similar. The UNSCETDG operates under the same procedural regime as the UNSCEGHS; thus, their composition and the decision-making processes are similar. NGOs participate in the work of the UNSCEGHS. Information about its activities can be obtained on the Internet, including decisions and agendas. Like GHS, UNRTDG are dynamic and are constantly being revised.

To conclude, the transnational bureaucracy networks and the transnational public law created by these networks are legitimized almost exclusively through alternative mechanisms. In all four cases – the IPCS, OECD Test Guidelines, GHS, and UNRTDG – the previously established requirements are fulfilled, which is rather surprising in view of the poor record of IFROs.

The requirement to give reasons is met in this case, because, in all four cases, decisions are based on the outcome of extensive scientific dialogue, represented by a continuous exchange of argument and counterargument and substantiated by current scientific understanding. Often the discussions take place in scientific journals.

The diverse composition of the committees is due to their multinational character. Unlike the national committees, no particular care is taken to ensure that experts with different views are admitted. Instead, in a similar way to the EU committees discussed above, conflicting views follow from the diverse risk cultures of the experts.

Each of the committees has its own website, publishing their decisions and agenda. This makes it relatively easy for the public to follow the process. NGOs are generally admitted, which not only ensures the representation of certain interests, but also ensures the transparency of the decision-making process.

Revisibility and currency also do not seem to be problematic issues. Particularly, the Test Guidelines, GHS, and UNRTDG are examples of how special care is taken in procedural rules to ensure standards are reviewed and reflect scientific progress.

In the end, the matter of values is the only one that gives reason for objection. The governing values passed down from the parent organization should guide the activities of the networks. However, in view of the apparent autonomy of the networks, they might stray from these original values. The transnational bureaucracy networks examined here all operate under rather strict and precise objectives laid down in their founding acts. With their growing autonomy, fears of a "global technocracy" are likely to rise.

18
Prospects

Despite the overall good performance of transnational bureaucracy networks and transnational public law in terms of legitimacy, there are still some things left to be desired. However, several things can be done, both on the national and the international level to rectify these issues.

The performance of the German *Bundestag* is poor with respect to transnational bureaucracy networks and transnational public law, and it is unlikely that other parliaments are fully aware of the issue of the devolution of lawmaking power to the transnational level either. Notwithstanding the usefulness of alternative legitimacy mechanisms and the fact that transnational public law is endorsed by Parliament, it should not be forgotten that the Parliament alone represents citizens and is supposed to be both the primary legislator and monitor of administrative activities.

Parliament must first become better aware of transnational relationships. Although not a new phenomenon, transnational relationships have shown intensified growth over the past decades. Increased awareness of this political process could result in Parliament employing mechanisms that it has used in the past to exercise control over the administration's transnational activities. Some scholars have suggested the possibility of establishing parliamentary networks or that parliamentarians accompany delegations; however, up to now, Parliament has not fully explored such ideas.

Second, the set of legal instruments available to Parliament could be expanded. In Germany, for example, the role of the Committee on Foreign Affairs could be readjusted to resemble that of the Committee on European Affairs, intensifying cooperation between the Federal Government and the Committee. This would include increased government awareness of administrative transnational activities. In addition, new legislation could be put in place to regulate the conduct of national agencies in transnational settings. Such a law would likely contribute to greater legal certainty in transnational environments. In Germany, it might be useful to revisit the proposed laws on standard setting in the *Umweltgesetzbuch – Entwurf der unabhängigen Sachverständigenkommission (UGB-KomE)* and the Commission on Risks.

A key challenge when implementing such legitimizing measures is to preserve the benefits of transnational bureaucracy networks and the laws they create. New laws and increased parliamentary supervision should not be used to stifle transnational relations.

At the international level, it might be helpful to establish a code of conduct for administrative activities to systematize the procedures assumed by transnational bureaucracies. Scholars have already observed the emergence of a global administrative law (Kingsbury, Krisch, and Stewart 2005), and identified certain "tools" used in this area (Esty 2006: 1523ff.).

There are two ways in which such law is formed. Under the "bottom-up" approach, the domestic administrative laws are expanded in scope and requirements altered to function at international and transnational levels.[377] Regulatory elements originating from various domestic legal orders might coalesce into transnational administrative law (Stewart 2005: 72). Alternatively, under a "top-down" approach international regimes are furnished with their own administrative structures and procedures (Id.: 76ff.). The transnational bureaucracy networks explored in this book all operate on the basis of MoU, RoP, or ToR, which lay down the procedures and satisfy basic legitimacy requirements. This implies a "top-down" approach, as the actual administrative activity, that is, creating transnational public law, is guided by these rules.

Yet, in view of the ad hoc, informal character of procedural rules laid down in MoUs, ToR, or RoP and the reluctance of states to give up sovereignty, it is likely that for the time being a "bottom-up" approach has greater appeal and prospects for success.

This book has demonstrated how legal orders intertwine. The findings suggest that the global administrative law will eventually emerge on the transnational level, where "bottom-up" and "top-down" approaches amalgamate.

---

[377] For the case of the US Administrative Procedure Act (APA) cf. Stewart 2005: 76ff.

Bibliography

Abbott, Kenneth W. and Snidal, Duncan (1998) "Why States Act through Formal International Organizations", *Journal of Conflict Resolution*, Vol. 42, pp. 3–32.

Abbott, Kenneth W. and Snidal, Duncan (2000) "Hard and Soft Law in International Governance", *International Organization*, Vol. 54, pp. 421–56.

Achterberg, Norbert (1984) *Parlamentsrecht* (Tübingen: Mohr Siebeck).

Adolphsen, Jens (2002) "Eine lex sportiva für den internationalen Sport?", *Jahrbuch Junger Zivilrechtswissenschaftler*, Vol. 11, pp. 281–301.

Alexy, Robert (1994) *Theorie der Grundrechte*, 2nd edn (Frankfurt am Main: Suhrkamp).

Alexy, Robert (2002) *Begriff und Bedeutung des Rechts* (Freiburg (Breisgau); München: Alber).

Alston, Philip (1978) "International Regulation of Toxic Chemicals", *Ecology Law Quarterly*, Vol. 7, pp. 397–456.

Amerasinghe, Chittharanjan Felix (2005) *Principles of the Institutional Law of International Organizations*, 2nd edn (Cambridge: Cambridge University Press).

American Chemistry Council (2003) *Comments of the American Chemistry Council on the European Commission's Proposal Concerning the Registration, Evaluation, Authorization and Restrictions of Chemicals* (Arlington, VA: American Chemistry Council).

Anders, Gerhard (2004) "Lawyers and Anthropologists – A Legal Pluralist Approach to Global Governance" in Dekker, Ige F. and Werner, Wouter G. (eds) *Governance and International Legal Theory* (Leiden; Boston, MA: Martinus Nijhoff), pp. 37–58.

Annan, Kofi (1999) "Address to the World Economic Forum in Davos, Switzerland, on 31 January 1999", *United Nations Press Release SG/SM/6881, 1 February 1999*, http://www.un.org/News/ossg/sg/stories/statments_search_full.asp?statID=22, date accessed 16 November 2008.

Archibugi, Daniele (1995) "From the United Nations to Cosmopolitan Democracy" in Held, David and Archibugi, Daniele (eds) *Cosmopolitan Democracy – An Agenda for a New World Order* (Oxford; Cambridge: Polity Press), pp. 121–62.

Arnull, Anthony, Dashwood, Alan, Ross, Malcolm, and Wyatt, Derrick (2000), *European Union Law*, 4th edn (London: Sweet & Maxwell).

Aust, Anthony (2000) *Modern Treaty Law and Practice* (Cambridge: Cambridge University Press).

Austin, John (1885) *Lectures on Jurisprudence or the Philosophy of Positive Law*, Vol. I, 5th edn, rev. (London: John Murray).

Badura, Peter (2001) "The Constitutional Basis of Public Administration" in König, Klaus and Siedentopf, Heinrich (eds), *Public Administration in Germany* (Baden-Baden: Nomos), pp. 47–56.

Badura, Peter (2003) *Staatsrecht – Systematische Erläuterung des Grundgesetzes*, 3rd edn, rev. (München: C. H. Beck).

Baldus, Manfred (1997) "Zur Relevanz des Souveränitätsproblems für die Wissenschaft vom öffentlichen Recht", *Der Staat*, Vol. 36, pp. 380–98.

Barnett, Michael and Finnemore, Martha (2004) *Rules for the World – International Organizations in Global Politics* (Ithaca, NY; London: Cornell University Press).

Beck, Ulrich (1997) *Was ist Globalisierung? –Irrtümer des Globalismus – Antworten auf Globalisierung*, 3rd edn (Frankfurt am Main: Suhrkamp).

Beck, Ulrich (2005) "World Risk Society and the Changing Foundations of Transnational Politics" in Grande, Edgar and Pauly, Louis W. (eds) *Complex Sovereigtny: Reconstituting Political Authority in the Twenty-first Century* (Toronto; Buffalo, NY; London: University of Toronto Press), pp. 22–47.

Beetham, David (1998) "Legitimacy" in Craig, Edward (ed.) *Routledge Encyclopedia of Philosophy*, Vol. 5 (London; New York: Routledge), pp. 538–41.

Benz, Arthur (1995) "Politiknetzwerke in der Horizontalen Politikverflechtung" in Jansen, Dorothea and Schubert, Klaus (eds) *Netzwerke und Politikproduktion – Konzepte, Methoden, Perspektiven* (Marburg: Schüren), pp. 185–204.

Berber, Friedrich (1975) *Lehrbuch des Völkerrechts*, Vol. 1, 2nd edn (München: C. H. Beck).

Bergström, Carl Fredrik (2005) *Comitology – Delegation of Powers in the European Union and the Committee System* (Oxford: Oxford University Press).

Bernárdez, Santiago Torres (1988) "Subsidiary Organs" in Dupuy, René-Jean (ed.) *Manuel sur les organisations internationals* (Dordrecht; Boston, MA: Martinus Nijhoff), pp. 100–46.

Beyerlin, Ulrich (2000) *Umweltvölkerrecht* (München: C. H. Beck).

Birnie, Patricia and Boyle, Alan (eds) (2002) *International Law and the Environment*, 2nd edn (Oxford: Oxford University Press).

Black, Donald (1976) *The Behavior of Law* (New York; San Francisco, CA; London: Academic Press).

Blanke, Thomas (1998) "Antidemokratische Effekte der verfassungsgerichtlichen Demokratietheorie", *Kritische Justiz*, Vol. 31, pp. 452–71.

Bleckmann, Albert (1995) *Staats- und Völkerrechtslehre – Vom Kompetenz- zum Kooperationsvölkerrecht* (Köln: Heymanns).

Böckenförde, Ernst-Wolfgang (2004) "Demokratie als Verfassungsprinzip" in Isensee, Josef and Kirchhof, Paul (eds) *Handbuch des Staatsrechts der Bundesrepublik Deutschland*, Vol. II, 3rd edn, rev. (Heidelberg: C. F. Müller), pp. 429–96.

Böckenförde, Ernst-Wolfgang (2005) "Demokratische Willensbildung und Repräsentation" in Isensee, Josef and Kirchhof, Paul (eds) *Handbuch des Staatsrechts der Bundesrepublik Deutschland*, Vol. III, 3rd edn, rev. (Heidelberg: C. F. Müller), pp. 31–54.

Bodansky, Daniel (1999) "The Legitimacy of International Governance – A Coming Challenge for International Environmental Law", *American Journal of International Law (AJIL)*, Vol. 93, pp. 596–624.

Bodin, Jean (1981) *Sechs Bücher über den Staat*, Vol. I (München: C. H. Beck).

Bohne, Eberhard (1981) *Der informale Rechtsstaat – Eine empirische und rechtliche Untersuchung zum Gesetzesvollzug unter besonderer Berücksichtigung des Immissionsschutzes* (Berlin: Duncker & Humblot).

Börzel, Tanja A. (1998) "Organizing Babylon – On the Different Conceptions of Policy Networks", *Public Administration*, Vol. 76, pp. 253–73.

Bradley, Kieran St Clair (1992) "Comitology and the Law – Through a Glass Darkly", *Common Market Law Review*, Vol. 29, pp. 693–721.

Brandsch, Romana, Nowak, Karl-Ernst, Binder, Norbert and Jastorff, Bernd (2002) "Untersuchungen zur Nachhaltigkeit der Sanierung von tributylzinnkontaminiertem Hafensediment durch Landablagerung", *Umweltwissenschaften und Schadstoff-Forschung – Zeitschrift für Umweltchemie und Ökotoxikologie,* Vol. 14, pp. 138–44.

Brody, Julia Green, Aschengrau, Ann, McKelvey, Wendy, Rudel, Ruthann A., Swartz, Christopher and Kennedy, Theresa (2004) "Breast Cancer Risk and Historical Exposure to Pesticides from Wide-Area Applications Assessed with GIS", *Environmental Health Perspectives,* Vol. 112, pp. 889–97.

Bröhmer, Jürgen (2004) *Transparenz als Verfassungsprinzip – Grundgesetz und Europäische Union* (Tübingen: Mohr Siebeck).

Brohm, Winfried (1987) "Sachverständige Beratung des Staates" in Isensee, Josef and Kirchhof, Paul (eds) *Handbuch des Staatsrechts der Bundesrepublik Deutschland,* Vol. II (Heidelberg: Müller), pp. 207–48.

Brownlie, Ian (2003) *Principles of Public International Law,* 6th edn (Oxford: Oxford University Press).

Brugger, Winfried (2000) "Gemeinwohl als Ziel von Staat und Recht" in Murswiek, Dietrich, Storost, Ulrich and Wolff, Heinrich A. (eds) *Staat – Souveränität – Verfassung: Festschrift für Helmut Quaritsch zum 70. Geburtstag* (Berlin: Duncker & Humblot), pp. 45–72.

Calliess, Gralf-Peter (2002) "Reflexive Transnational Law – The Privatisation of Civil Law and the Civilisation of Private Law", *Zeitschrift für Rechtssoziologie,* Vol. 23, pp. 185–216.

Carpenter, Chad and Krueger, Jonathan (1997) "A Brief History of the IFCS", Earth Negotiations Bulletin, 17 February 1997, Vol. 15, pp. 1–3.

Cassese, Antonio (1985) "Modern Constitutions and International Law", *Recueil des Cours,* Vol. 192, pp. 331–473.

Cassese, Antonio (1986) *International Law in a Divided World* (Oxford: Clarendon).

Cassese, Sabino (1985) "Die Beziehungen zwischen den internationalen Organisationen und den nationalen Verwaltungen" in Achour, Yadh Ben and Cassese, Sabino (eds) *Aspekte der internationalen Verwaltung* (Baden-Baden: Nomos), pp. 37–75.

CEFIC (2004) *Chemistry – Elements of Life* (Brussels: CEFIC).

CEFIC (2008a) *Chemistry and You,* http://www.chemistryandyou.org, date accessed 16 November 2008.

CEFIC (2008b) *Elements of Life,* http://www.elements-of-life.org, date accessed 16 November 2008.

CEFIC (2008c) *Establishing and Working in International Consortia,* http://www.cefic.org/activities/hse/mgt/hpv/hpvidescr-app1.htm, date accessed 16 November 2008.

CEFIC (2008d) *Welcome to the Website of the Global Initiative on High Production Volume (HPV) Chemicals!* http://www.cefic.org/activities/hse/mgt/hpv/hpvinit.htm, date accessed 16 November 2008.

Champ, Michael A. (2000) "A Review of Organotin Regulatory Strategies, Pending Actions, Related Costs and Benefits", *The Science of the Total Environment,* Vol. 258, pp. 21–71.

Chayes, Abram and Chayes, Antonia Handler (1995) *The New Sovereignty – Compliance With International Regulatory Agreements* (Cambridge, MA: Harvard University Press).

Chemical Abstract Service (2008a) *CAS REGISTRY*, http://www.cas.org/expertise/cascontent/registry/index.html, date accessed 16 November 2008.

Chemical Abstract Service (2008b) *CHEMCATS*, http://www.cas.org/expertise/cascontent/chemcats.html, date accessed 16 November 2008.

Claude, Inis L. (1960) *Swords into Plowshares – The Problems and Progress of Intenational Organizations*, 3rd edn, rev. (New York: Random House).

Commission on Global Governance (1995) *Our Global Neighbourhood – The Report of the Commission on Global Governance* (Oxford: Oxford University Press).

Czempiel, Ernst-Otto (1995) "Aktivieren, negieren, reformieren? – Zum 50jährigen Bestehen der Vereinten Nationen", *Aus Politik und Zeitgeschichte*, Vol. 42, pp. 36–45.

Dahm, Georg, Delbrück, Jost and Wolfrum, Rüdiger (1989) *Völkerrecht*, Vol. I/1, 2nd edn, rev. (Berlin and New York: de Gruyter).

Dahm, Georg, Delbrück, Jost and Wolfrum, Rüdiger (2002) *Völkerrecht*, Vol. I/2, 2nd edn, rev. (Berlin and New York: de Gruyter, 2002).

De Marcellus, Sally (2003) "OECD Environment, Health and Safety Programme – Information on the World Wide Web", *Toxicology*, Vol. 190, pp. 125–34.

De Vattel, Emer (1959) *Le Droit Des Gens Ou Principes De La Loi Naturelle – Appliques à La Conduite Et Aux Affaires Des Nations Et De Souverains* (Tübingen: J. C. B. Mohr (Paul Siebeck)).

Dehousse, Renaud (2003) "Comitology – Who watches the watchmen?", *Journal of European Public Policy*, Vol. 10, pp. 793–813.

Delbrück, Jost (1987) "Internationale und nationale Verwaltung – Inhaltliche und institutionelle Aspekte" in Geserich, Kurt G. A. and Pohl, Hans (eds) *Deutsche Verwaltungsgeschichte*, Vol. 5 (Stuttgart: DVA), pp. 386–403.

Delbrück, Jost (1993) "A More Effective International Law or a New 'World Law'? – Some Aspects of the Development of International Law in a Changing International System", *Indiana Law Journal*, Vol. 68, pp. 705–25.

Delbrück, Jost (2001) "Structural Changes in the International System and Its Legal Order – International Law in the Era of Globalization", *Schweizerische Zeitschrift für internationales und europäisches Recht*, Vol. 11, pp. 1–36.

Delbrück, Jost (2002a) "Das Staatsbild im Zeitalter wirtschaftsrechtlicher Globalisierung", *Arbeitspapiere aus dem Institut für Wirtschaftsrecht*, No. 3.

Delbrück, Jost (2002b) "Prospects for a 'World (Internal) Law' – Legal Developments in a Changing International System", *Indiana Journal for Global Legal Studies*, Vol. 9, pp. 401–31.

Delbrück, Jost (2003) "Exercising Public Authority Beyond the State – Transnational Democracy and/or Alternative Legitimation Strategies?", *Indiana Journal for Global Legal Studies*, Vol. 10, pp. 29–43.

Denninger, Erhard (1990) *Verfassungsrechtliche Anforderungen an die Normsetzung im Umwelt- und Technikrecht* (Baden-Baden: Nomos).

di Fabio, Udo (2003) "Verfassungsstaat und Weltgesellschaft" in Mellinghoff, Rudolf, Morgenthaler, Gerd and Puhl, Thomas (eds) *Die Erneuerung des Verfassungsstaates – Symposium aus Anlass des 60. Geburtstages von Professor Dr. Paul Kirchhof* (Heidelberg: C. F. Müller), pp. 57–79.

Dicke, Klaus (1994) *Effizienz und Effektivität Internationaler Organisationen – Darstellung und kritische Analyse eines Topos im Reformprozeß der Vereinten Nationen* (Berlin: Duncker & Humblot).

Diderich, Robert (2007) "The OECD Chemicals Programme" in van Leeuwen, Cornelis Johannes and Vermeire, Theodorus Gabriël (eds) *Risk Assessment of Chemicals: An Introduction* (Dordrecht: Springer), pp. 623–38.

Dreier, Ralf (1986) "Der Begriff des Rechts", *Neue Juristische Wochenschrift*, Vol. 39, pp. 890–6.

Dupuy, Pierre-Marie (1991) "Soft Law and the International Law of the Environment", *Michigan Journal of International Law*, Vol. 12, pp. 420–34.

Ebbecke, Katja, "Giftiger Reimport", *Süddeutsche Zeitung*, 23 March 2006, 18.

Eberwein, Wolf-Dieter and Kaiser, Karl (1998) "Wissenschaft und außenpolitischer Entscheidungsprozeß" in Eberwein, Wolf-Dieter and Kaiser, Karl (eds) *Deutschlands neue Außenpolitik*, Vol. 4 (München: Oldenbourg), pp. 1–12.

EDF (1997) *Toxic Ignorance – The Continuing Absence of Basic Health Testing for Top-Selling Chemicals in the United States*, http://www.edf.org/documents/243_toxicignorance.pdf, date accessed 16 November 2008.

Ehrlich, Eugen (1989) *Grundlegung der Soziologie des Recht*, 4th edn (Berlin: Duncker & Humblot).

Emory, Richard W. (2000) "Probing the Protections in the Rotterdam Convention on Prior Informed Consent", *Colorado Journal of International Environmental Law and Policy*, Vol. 12, pp. 47–69.

EPA (1972) "DDT Ban Takes Effect", *EPA Press Release*, 31 December 1972, http://www.epa.gov/history/topics/ddt/01.htm, date accessed 16 November 2008.

EPA (1998a) *Chemical Hazard Data Availability Study – What Do We Really Know About the Safety of High Production Volume Chemicals?* http://www.epa.gov/chemrtk/pubs/general/hazchem.pdf, date accessed 16 November 2008.

EPA (1998b) *Letter to Manufacturers/Importers*, http://www.epa.gov/HPV/pubs/general/ceoltr1.htm, date accessed 16 November 2008.

Esty, Daniel C. (2006) "Good Governance at the Supranational Scale: Globalizing Administrative Law", *Yale Law Journal*, Vol. 115, pp. 1490–562.

European Commission (1999) *Public Availability of Data on EU High Production Volume Chemicals*, http://ecb.jrc.ec.europa.eu/Data-Availability-Documents/datavail.pdf, date accessed 16 November 2008.

European Commission (2008) *Comitology Register*, http://ec.europa.eu/transparency/regcomitology, date accessed 16 November 2008.

European Workshop on the Impact of Endocrine Disruptors on Human Health and Wildlife (1996) *Report of the Proceedings* (Weybridge: European Commission).

Falke, Josef (1995) "Informationspolitische Maßnahmen im Chemikalienrecht" in Winter, Gerd (ed.) *Risikoanalyse und Risikoabwehr im Chemikalienrecht – Interdisziplinäre Untersuchungen* (Düsseldorf: Werner), pp. 65–114.

Falke, Josef (1996) "Comitology and Other Committees – A Preliminary Empirical Assessment" in Pedler, Robin H. and Schäfer, Günther F. (eds) *Shaping European Law and Policy: The Role of Committees and Comitology in the Political Process* (Maastrich: European Institute of Public Administration), pp. 117–65.

Falke, Josef (2000) *Rechtliche Aspekte der Normung in den EG-Mitgliedstaaten und der EFTA*, Vol. 3 (Luxemburg: Amt für amtliche Veröffentlichungen der Europäischen Gemeinschaften).

Falke, Josef and Winter, Gerd (1996) "Management and Regulatory Committees in Executive Rule-Making" in Winter, Gerd (ed.) *Sources and Categories in European Union Law – A Comparative and Reform Perspective* (Baden-Baden: Nomos), pp. 541–82.

FAO (2008) *Welcome to FAO Pesticide Management*, http://www.fao.org/ag/AGP/agpp/Pesticid/default.htm, date accessed 16 November 2008.

Farazmand, Ali (1999) "Globalization and Public Administration", *Public Administration Review*, Vol. 59, pp. 509–22.

Fassbender, Bardo (1998) "Die verfassungs- und völkerrechtsgeschichtliche Bedeutung des Westfälischen Friedens von 1648" in Erberich, Ingo (ed.) *Frieden und Recht – 38. Tagung der Wissenschaftlichen Mitarbeiterinnen und Mitarbeiter der Fachrichtung "Öffentliches Recht", Münster 1998* (Stuttgart: Boorberg), pp. 9–52.

Feller, Silke, Kowalski, Ulrike and Schlottmann, Ulrich (2005) *National Profile – Chemikalienmanagement in Deutschland*, 2nd edn, rev. (Dortmund: Bundesanstalt für Arbeitsschutz und Arbeitsmedizin).

Fidler, David P. and Gostin, Lawrence O. (2006) "The New International Health Regulations – An Historic Development for International Law and Public Health", *Journal of Law, Medicine and Ethics*, Vol. 34, pp. 85–96.

Fiedler, Heidelore (2003) "Introduction" in Fiedler, Heidelore (ed.) *Persistent Organic Pollutants* (Berlin: Springer), pp. XI–XII.

Fisahn, Andreas (2002) *Demokratie und Öffentlichkeitsbeteiligung* (Tübingen: Mohr Siebeck).

Fischer, Wolfgang and Holtrup, Petra (1998) "Institutionelle Strukturen und Entscheidungsprozesse der Umweltaußenpolitik" in Eberwein, Wolf-Dieter and Kaiser, Karl (eds) *Deutschlands neue Außenpolitik*, Vol. 4 (München: Oldenbourg), pp. 121–36.

Fischer-Lescano, Andreas and Teubner, Gunther (2004) "Regime-Collisions – The Vain Search for Legal Unity in the Fragmentation of Global Law", *Michigan Journal of International Law*, Vol. 25, pp. 999–1045.

Forsthoff, Ernst (1973) *Lehrbuch des Verwaltungsrechts*, Vol. 1, 10th edn, rev. (München: C. H. Beck).

Fortune (2008) *Global 500*, http://money.cnn.com/magazines/fortune/global500/2008/full_list, date accessed 16 November 2008.

Foster, Nigel G. and Sule, Satish (2002) *German Legal System & Laws*, 3rd edn (Oxford and New York: Oxford University Press).

Freiherr von Lersner, Heinrich (1990) "Verfahrensvorschläge für umweltrechtliche Grenzwerte", *Natur und Recht*, Vol. 12, pp. 193–7.

Friedman, Lawrence M. (1996) "Borders: On the Emerging Sociology of Transnational Law", *Stanford Journal of International Law*, Vol. 32, pp. 65–90.

Friedmann, Wolfgang (1994) *The Changing Structure of International Law* (London: Stevens & Sons).

Frowein, Jochen A. (2000) "Konstitutionalisierung des Völkerrechts" in Deutsche Gesellschaft für Völkerrecht (ed.) *Völkerrecht und Internationales Privatrecht in einem sich globalisierenden internationalen System* (Heidelberg: C. F. Müller), pp. 427–45.

Führ, Martin (1993) "Technische Normen in demokratischer Gesellschaft – Zur Steuerung der Technikanwendung durch private Normen", *Zeitschrift für Umweltrecht*, Vol. 5, pp. 99–102.

Gärtner, Sabine, Küllmer, Jens and Schlottmann, Ulrich (2003) "Chemikaliensicherheit in einer verletzlichen Welt", *Angewandte Chemie*, Vol. 115, pp. 4594–607.

Geiger, Rudolf (2002) *Grundgesetz und Völkerrecht – Die Bezüge des Staatsrechts zum Völkerecht und Europarecht*, 3rd edn, rev. (München: C. H. Beck).

Gessner, Volkmar (2002) "Rechtspluralismus und globale soziale Bewegungen", *Zeitschrift für Rechtssoziologie*, Vol. 23, pp. 277–305.

Ginzky, Harald (2000) "Vermarktungs- oder Verwendungsbeschränkungen von Chemikalien–Verfahren, materielle Anforderungen und Reformüberlegungen", *Zeitschrift für Umweltrecht*, Vol. 11, pp. 129–36.

Graf Kielmansegg, Peter (1996) "Integration und Demokratie", in Jachtenfuchs, Markus and Kohler-Koch, Beate (eds) *Europäische Integration* (Opladen: Leske + Budrich), pp. 47–71.

Grams, Hartmut A. (1998) *Zur Gesetzgebung der Europäischen Union – Eine vergleichende Strukturanalyse aus staatsorganisatorischer Sicht* (Neuwied and Kriftel: Luchterhand).

Gray, Mark Allan (1990) "The United Nations Environment Programme – An Assessment", *Environmental Law*, Vol. 20, pp. 291–318.

Green, Leslie (1998) "Authority" in Craig, Edward (ed.) *Routledge Encyclopedia of Philosophy*, Vol. 1 (London; New York: Routledge) pp. 584–6.

Grewe, Wilhelm G. (2000) *The Epochs of International Law* (Berlin and New York: Walter de Gruyter).

Grewe, Wolfgang G. (1988) "Auswärtige Gewalt" in Isensee, Josef and Kirchhof, Paul (eds) *Handbuch des Staatsrechts der Bundesrepublik Deutschland*, Vol. III (Heidelberg: Müller), pp. 921–75.

Gross, Leo (1948) "The Peace of Westphalia, 1648–1948", *American Journal of International Law*, Vol. 42, pp. 20–41.

Grotius, Hugo (1950) *De Jure Bellis Ac Pacis Libri Tres* (Tübingen: J. C. B. Mohr (Paul Siebeck)).

Grünfeld, H. T. and Bonefeld-Jorgensen E. C. (2004) "Effect of in vitro Estrogenic Pesticides on Human Oestrogen Recepter α and β mRNA Levels", *Toxicology Letters*, Vol. 151, pp. 467–80.

Günther, Klaus (2001) "Rechtspluralismus und universaler Code der Legalität – Globalisierung als rechtstheoretisches Problem" in Wickert, Lutz and Günther, Klaus (eds) *Die Öffentlichkeit der Vernunft und Vernunft der Öffentlichkeit – Festschrift für Jürgen Habermas* (Frankfurt am Main: Suhrkamp), pp. 539–67.

Gusy, Christoph (1986) "Wertungen und Interessen in der technischen Normung", *Umwelt- und Planungsrecht*, Vol. 5, pp. 241–50.

Gusy, Christoph (1994) "Die untergesetzliche Rechtsetzung nach dem Bundesimmissionsschutzgesetz aus verfassungsrechtlicher Sicht", in Koch, Hans-Joachim and Lechelt, Rainer (eds) *Zwanzig Jahre Bundes-Immissionsschutzgesetz* (Baden-Baden: Nomos), pp. 185–216.

Gusy, Christoph (1995) "Probleme der Verrechtlichung technischer Standards", *Neue Juristische Wochenschrift*, Vol. 48, pp. 105–12.

Häberle, Peter (1978) "Der kooperative Verfassungsstaat" in Kaulbach, Friedrich and Krawietz, Werner (eds) *Recht und Gesellschaft: Festschrift für Helmut Schelsky zum 65. Geburtstag* (Berlin: Duncker & Humblot), pp. 141–77.

Haenel, Albert (1892) *Deutsches Staatsrecht*, Vol. 1 (Leipzig: Duncker & Humblot).

Hahn, Hugo J. (1997) "Organisation for Economic Co-Operation and Development", in Bernhardt, Rudolf (ed.) *Encyclopedia Public International Law*, Vol. III (Amsterdam: North-Holland), pp. 791–9.

Hailbronner, Kay (1997) "Kontrolle der auswärtigen Gewalt", *Veröffentlichungen der Vereinigung der Deutschen Staatsrechtslehrer*, Vol. 56, pp. 7–37.

Harris, Catherine A. and Sumpter, John P. (2001) "The Endocrine Disrupting Potential of Phtalates" in Metzler, Manfred (ed.) *Endocrine Disruptors*, Part I (Berlin: Springer), pp. 169–201.

Hart, Herbert Lionel Adolphus (1994) *The Concept of Law*, 2nd edn (Oxford: Clarendon Press).

Hartley, Trevor C. (2003) *The Foundations of European Law – An Introduction to the Constitutional and Administrative Law of the European Community*, 5th edn (Oxford: Oxford University Press).

Hartley, Trevor C. (2004) *European Union in a Global Context – Text, Cases and Materials* (Cambridge: Cambridge University Press).

Hell, Karl-Heinz (1990) "Die Auswirkungen des Sandoz-Unfalles auf die Biozönose des Rheins" in Kinzelbach, Ragnar and Friedrich, Günther (eds) *Biologie des Rheins* (Stuttgart: Fischer), pp. 11–26.

Held, David (1995) "Democracy and the New International Order" in Held, David and Archibugi, Daniele (eds) *Cosmopolitan Democracy – An Agenda for a New World Order* (Oxford and Cambridge: Polity Press), pp. 96–120.

Held, David, McGrew, Anthony G., Goldblatt, David and Perraton, Jonathan (2003) *Global Transformations – Politics, Economics and Culture* (Oxford: Polity Press).

Henkin, Louis (1989) "International Law: Politics, Values, Functions – General Course on Public International Law", *Recueil des Cours*, Vol. 216, pp. 9–416.

Henkin, Louis (1997) *Foreign Affairs and the United States Constitution*, 2nd edn (Oxford: Clarendon Press).

Herberg, Martin (2005) "Entkoppeltes Recht? –Die Umweltstandards multinationaler Konzerne zwischen Informalität und Verrechtlichung", *TranState Working Papers*, No. 20.

Herberg, Martin (2006) "Private Authority, Global Governance, and the Law – The Case of Environmental Self-Regulation in Multinational Enterprises" in Winter, Gerd (ed.) *Multilevel Governance of Global Environmental Change – Perspectives from Science, Sociology and the Law* (Cambridge: Cambridge University Press), pp. 149–78.

Herdegen, Matthias (2003) *Internationales Wirtschaftsrecht*, 4th edn (München: C. H. Beck).

Hertel, Rolf (2004) "Zweck von Risikomanagement" in Reichl, Franz-Xaver and Schwenk, Michael (eds) *Regulatorische Toxikologie* (Berlin: Springer), pp. 428–37.

Hesse, Konrad (1999) *Grundzüge des Verfassungsrechts der Bundesrepublik Deutschland*, reprint of the 20th edn (Heidelberg: C. F. Müller).

Hildebrandt, Bernd-Ulrich and Schlottmann, Ulrich (1998) "Chemikaliensicherheit eine internationale Herausforderung", *Angewandte Chemie*, Vol. 110, pp. 1383–93.

Hillgruber, Christian (2004) "Der Nationalstaat in der überstaatlichen Verflechtung" in Isensee, Josef and Kirchhof, Paul (eds) *Handbuch des Staatsrechts der Bundesrepublik Deutschland*, Vol. II, 3rd edn (Heidelberg: Müller), pp. 929–92.

Hinds, Caroline (1997) *Umweltrechtliche Einschränkungen der Souveränität – Völkerrechtliche Präventionspflicht zur Verhinderung von Umweltschäden* (Frankfurt: Lang). Hingst, Ulla (2001) *Auswirkungen der Globalisierung auf das Recht der Völkerrechtlichen Verträge* (Berlin: Duncker & Humblot).

Hiraishi, Takahiko (1989) "Control of Chemicals" in Tsuro, Shigeto and Weidner, Helmut (eds) *Environmental Policy in Japan* (Berlin: Ed. Sigma Bohn), pp. 332–41.

Hobbes, Thomas (1991) *Leviathan* (Cambridge: Cambridge University Press).

Hobe, Stephan (1997) "Global Challenges to Statehood – The Increasingly Important Role of Nongovernmental Organizations", *Indiana Journal for Global Legal Studies*, Vol. 5, pp. 191–208.

Hobe, Stephan (1998a) "Der kooperationsoffene Verfassungsstaat", Der Staat, Vol. 36, pp. 521–46.

Hobe, Stephan (1998b) *Der offene Verfassungsstaat zwischen Souveränität und Interdependenz – Eine Studie zur Wandlung des Staatsbegriffs der deutschsprachigen Staatslehre im Kontext internationaler internationalisierter Kooperation* (Berlin: Duncker & Humblot).

Hobe, Stephan (1999a) "Der Rechtsstatus der Nichtregierungsorganisationen", *Archiv des Völkerrechts*, Vol. 37, pp. 152–76.

Hobe, Stephan (1999b) "Die Zukunft des Völkerrechts im Zeitalter der Globalisierung", *Archiv des Völkerrechts*, Vol. 37, pp. 253–82.

Hobe, Stephan (2002) "Globalisation – A Challenge to the Nation State and International Law" in Likosky, Michael (ed.) *Transnational Legal Process: Globalisation and Power Disparities* (London: Butterworths), pp. 378–91.

Hobe, Stephan and Kimminich, Otto (2004) *Einführung in das Völkerrecht*, 8th edn, rev. (Tübingen: Francke).

Hoebel, E. Adamson (1968) *Das Recht der Naturvölker – Eine vergleichende Untersuchung rechtlicher Abläufe* (Olten; Freiburg im Breisgau: Walter Verlag).

Höreth, Marcus (1998) *The Trilemma of Legitimacy – Multilevel Governance in the EU and the Problem of Democracy* (Bonn: Zentrum für Europäische Integrationsforschung).

Huismans, J. W. (1980) "The International Register of Potentially Toxic Chemicals (IRPTC)", *Ecotoxicology and Environmental Safety*, Vol. 4, pp. 393–403.

Hummer, Waldemar (1998) "Die Beteiligung des Europäischen Parlaments an der 'Komitologie'" in Hafner, Georg, Loibl, Gerhard, Rest, Alfred, Sucharipa-Behrmann, Lilly and Zemanek, Karl (eds) *Liber Amicorum Professor Seidl-Hohenveldern in Honour of His 80th Birthday* (The Hague: Kluwer Law International), pp. 285–324.

ICCA (2005) *HPV Working List – Status of 28 April 2005*, http://www.cefic.be/ activities/hse/mgt/hpv/ICCA%20Working%20List%20-%20October%20 2005.xls, date accessed 16 November 2008.

ICCA (2008) *Alphabetic List of Consortia*, http://www.iccahpv.com/reports/ ReportConsortiaList.cfm, date accessed 16 November 2008.

ILO (2008a) *Areas of Harmonization*, http://www.ilo.org/public/english/protection/ safework/ghs/areasof.htm, date accessed 16 November 2008.

ILO (2008b) *Coordination Group for the Harmonization and Classification of Chemicals (CG/HCCS)*, http://www.ilo.org/public/english/protection/safework/ ghs/cghccs.htm, date accessed 16 November 2008.

ILO (2008c) *ILOLEX*, http://www.ilo.org/ilolex/cgi-lex/ratifce.pl?C170, date accessed 16 November 2008.

International Chamber of Commerce (1999) *Incoterms 2000 – ICC Official Rules for the Interpretation of Trade Terms* (Paris: ICC).

IOMC (2008a) *What is the IOMC?*, Brochure, http://www.who.int/iomc/IOMC_ brochure_en.pdf, date accessed 16 November 2008.

IOMC (2008b), *Technical Co-ordinating Groups,* http://www.who.int/iomc/groups/ en, date accessed 16 November 2008.

IPCS (2002) *Guidelines for the Preparation of Concise International Chemical Assessment Documents (CICADs),* http://www.who.int/ipcs/publications/cicad/ en/CICAD_Guidelines_sept-2002.pdf, date accessed 16 November 2008.

IPCS (2003) *Re-Designing IPCS – Current Situation* (Geneva: IPCS).

IPCS (2008a) *INCHEM Database,* http://www.inchem.org/pages/cicads.html, date accessed 16 November 2008.

IPCS (2008b) *Final Review Board and Consultative Group Meeting Reports,* http:// www.who.int/ipcs/publications/cicad/board_meeting_reports/en/index. html, date accessed 16 November 2008.

IPCS (2008c) *"Rules of Thumb" for Helping to Avoid Duplicative Activity Between IPCS and OECD Assessment Programmes,* http://www.who.int/ipcs/publications/ cicad/en/rules_of_thumb.pdf, date accessed 16 November 2008.

IPCS Risk Assessment Steering Group (2008) *Terms of Reference,* http://www.who. int/ipcs/publications/rasg/en/rasg-tr.pdf, date accessed 16 November 2008.

Ipsen, Knut (2004) *Völkerrecht,* 5th edn, rev. (München: C. H. Beck).

Jarass, Hans D. (1999) "Bindungswirkung von Verwaltungsvorschriften", *Juristische Schulung,* Vol. 39, pp. 105–12.

Jessup, Philip C. (1956) *Transnational Law* (New Haven, CT: Yale University Press).

Johnson, Lori A., Fujie, Tatsuya and Aalders, Marius (2000) "New Chemical Notification Laws in Japan, the United States, and the European Union" in Kagan, Robert A. and Axelrad, Lee (eds) *Regulatory Encounters – Multinational Corporations and American Adversarial Legalism* (Berkeley: University of California Press), pp. 341–71.

Jones, Wordsworth Filo and Yeater, Marcel D. (1992) "Hazardous Substances in Sand", in Peter H. (ed.) *The Effectiveness of International Environmental Agreements A Survey of Existing Legal Instruments* (Cambridge: Cambridge University Press), pp. 309–38.

Junne, Gerd C. A. (2001) "International Organizations in a Period of Globalization – New (Problems of) Legitimacy" in Coicaud, Jean-Marc and Heiskanen, Veijo (eds) *The Legitimacy of International Organizations* (Tokyo: United Nations University Press), pp. 189–220.

Kadelbach, Stefan (1992) *Zwingendes Völkerrecht* (Berlin: Duncker & Humblot).

Kadelbach, Stefan (2003) "Die parlamentarische Kontrolle des Regierungsh- andelns bei der Beschlußfassung in internationalen Organisationen" in Geiger, Rudolf (ed.) *Neuere Probleme der parlamentarischen Legitimation im Bereich der auswärtigen Gewalt* (Baden-Baden: Nomos), pp. 41–57.

Kadelbach, Stefan and Guntermann, Ute (2001) "Vertragsgewalt und Parlamentsvorbehalt – Die Mitwirkungsrechte des Bundestages bei sogen- annten Parallelabkommen und die völkerrechtliche Konsequenz ihrer Verletzung", *Archiv des öffentlichen Rechts,* Vol. 126, pp. 563–87.

Kaiser, Karl (1971) "Transnational Relations as a Threat to the Democratic Process", *International Organization,* Vol. 25, pp. 706–20.

Kaiser, Karl (1998) "Globalisierung als Problem der Demokratie", *Internationale Politik,* Vol. 53, pp. 3–11.

Kallenborn, Roland and Herzke, Dorte (2001) "Schadstoff-Ferntransport in die Arktis" *Umweltwissenschaften und Schadstoff-Forschung Zeitschrift für Umweltchemie und Ökotoxikologie,* Vol. 13, pp. 216–26.

Kamminga, Menno T. (2002) "The Evolving Status of NGOs under International Law – A Threat to the Inter-State System?" in Kreijen, Gerard (ed.) *State, Sovereignty, and International Governance* (Oxford: Oxford University Press), pp. 387–406.

Karkainnen, Bradley C. (2002) "Post-Sovereign Environmental Governance – The Collaborative Problem-Solving Model" in Biermann, Frank, Brohm, Rainer and Dingwerth, Klaus (eds) *"Global Environmental Change and the Nation State": Proceedings of the 2001 Berlin Conference on the Human Dimensions of Global Environmental Change* (Potsdam: Potsdam Institute for Climate Impact Research), pp. 206–16.

Keita-Ouane, Fatoumata, Durkee, Linda, Clevenstine, Emmert, Ruse, Michael, Csizer, Zoltan, Kearns, Peter and Halpaap, Achim (2001) "The IOMC Organizations – A Source of Chemical Safety Information", *Toxicology*, Vol. 157, pp. 111–19.

Kelsen, Hans (1944) "The Principle of Sovereign Equality of States as a Basis for International Organization", *Yale Law Journal*, Vol. 53, pp. 207–27.

Kelsen, Hans (1960), *Reine Rechtslehre*, 2nd edn, rev. (Wien: Franz Deuticke).

Kelsen, Hans (1966) *Allgemeine Staatslehre*, reprint of the 1st edn of 1925, (Bad Homburg: Verlag Dr. Max Gehlen).

Kelsen, Hans (1967) *Principles of International Law*, 2nd edn (New York: Holt, Rinehart, Winston).

Keohane, Robert O. and Nye, Joseph S. (1974) "Transgovernmental Relations and International Organizations", *World Politics*, Vol. 27, pp. 39–62.

Kilian, Michael (1987) *Umweltschutz durch Internationale Organisationen – Die Antwort des Völkerrechts auf die Krise der Umwelt?* (Berlin: Duncker & Humblot).

Kilian, Michael (1995) "UNEP – United Nations Environment Programme" in Wolfrum, Rüdiger (ed.) *United Nations – Law, Policies and Practice* (München and Dordrecht: C. H. Beck; Nijhoff), pp. 1296–304.

Kimminich, Otto (1976) "Gegenwärtiger Stand und Entwicklungstendenzen in der Souveränitätslehre und Souveränitätspraxis", in Kroker, Eduard and Veier, Theodor (eds) *Rechtspositivismus, Menschenrechte und Souveränitätslehre in verschiedenen Rechtskreisen* (Wien: Braumüller), pp. 97–109.

Kingsbury, Benedict, Krisch, Nico and Stewart, Richard B. (2005) "The Emergence of Global Administrative Law", *Law and Contemporary Problems*, Vol. 68, pp. 15–61.

Kirchhof, Ferdinand (1987) *Private Rechtsetzung* (Berlin: Duncker & Humblot).

Kistenbrügger, Lothar (2004) "The ICCA HPV Chemicals Initiative", *Zeitschrift für Stoffrecht*, Vol. 1, pp. 36–40.

Kjær, Anne Mette (2004) *Governance* (Cambridge: Polity Press).

Klabbers, Jan (1996) "The Redundancy of Soft Law", *Nordic Journal of International Law*, Vol. 65, pp. 167–82.

Klabbers, Jan (1998) "The Undesirability of Soft Law", *Nordic Journal of International Law*, Vol. 67, pp. 381–91.

Klabbers, Jan (2002) *An Introduction to International Institutional Law* (Cambridge: Cambridge University Press).

Klaschka, Ursula, Lange, Arno W. and Madle, Stephan (1997) "Das OECD-Prüfrichtlinienprogramm", *Umweltwissenschaften und Schadstoff – Forschung Zeitschrift für Umweltchemie und Ökotoxikologie*, Vol. 9, pp. 387–96.

Kloepfer, Michael (2003) "Instrumente des Technikrechts", in Schulte, Martin (ed.) *Handbuch des Technikrechts* (Berlin: Springer), pp. 111–53.

Köck, Heribert Franz and Fischer, Peter (1997) *Das Recht der Internationalen Organisationen – Das Recht der Weltfriedensorganisation (VB und UNO), der UN-Spezialorganisationen und der wichtigsten regionalen Organisationen im Dienste von kollektiver Sicherheit und Zusammenarbeit* 3rd edn (Wien: Linde).

Koëter, Herman B. W. M. (2003) "Mutual Acceptance of Data – Harmonised Test Methods and Quality Assurance of Data – The Process Explained", *Toxicology Letters*, Vols. 140–1, pp. 11–20.

Kokott, Juliane (1996) "Kontrolle der auswärtigen Gewalt", *Deutsches Verwaltungsblatt*, Vol. 111, pp. 937–50.

Kokott, Juliane (2003) "Die Staatsrechtslehre und die Veränderung ihres Gegenstandes – Konsequenzen von Europäisierung und Internationalisierung" *Veröffentlichungen der Vereinigung der Deutschen Staatsrechtslehrer*, Vol. 63, pp. 7–40.

Kommers, Donald P. (1997) *The Constitutional Jurisprudence of the Federal Republic of Germany*, 2nd edn (Durham and London: Duke University Press).

König, Klaus (2000) "Öffentliche Verwaltung und Globalisierung", *Verwaltungsarchiv*, Vol. 92, pp. 475–506.

Kopp, Manfred and Müller, Hans-Peter (1980) *Herrschaft und Legitimität in modernen Industriegesellschaften – Eine Untersuchung der Ansätze von Max Weber, Niklas Luhmann, Claus Offe, Jürgen Habermas* (München: tuduv Verlagsgesellschaft).

Korn, Horst (2004) "Institutioneller und instrumentaler Rahmen für die Erhaltung der Biodiversität" in Wolff, Nina and Köck, Wolfgang (eds) *10 Jahre Übereinkommen über biologische Vielfalt – Eine Zwischenbilanz* (Baden-Baden: Nomos), pp. 36–54.

Krause, Joachim (1998) "Die Rolle des Bundestages in der Außenpolitik" in Eberwein, Wolf-Dieter and Kaiser, Karl (eds) *Deutschlands neue Außenpolitik*, Vol. 4 (München: Oldenbourg), pp. 137–52.

Kunig, Philip (1989) "Deutsches Verwaltungshandeln und Empfehlungen internationaler Organisationen" in Hailbronner, Kay (ed.) *Staat- und Völkerrechtsordnung: Festschrift für Karl Doehring* (Berlin: Springer), pp. 529–51.

Lagoni, Rainer (1995) "ECOSOC – Economic and Social Council" in Wolfrum, Rüdiger (ed.) *United Nations – Law, Policies and Practice* (München and Dordrecht: C. H. Beck; Nijhoff), pp. 461–9.

Lahl, Uwe and Tickner, Joel (2004) "Defizite im amerikanischen und europäischen Chemikalienrecht – Reformbemühungen um die transatlantische Dialogbereitschaft", *Zeitschrift für Stoffrecht*, Vol. 1, pp. 156–70.

Lallas, Peter L. (2001) "The Stockholm Convention on Persistent Organic Pollutants", *American Journal of International Law*, Vol. 95, pp. 692–708.

Lampe, Ernst-Joachim (1995) "Was ist 'Rechtspluralismus'?" in Lampe, Ernst-Joachim (ed.) *Rechtsgleichheit und Rechtspluralismus* (Baden-Baden: Nomos), pp. 8–33.

Loewenstein, Karl (1954) "Sovereignty and International Co-operation", *American Journal of International Law*, Vol. 48, pp. 222–43.

Lowe, Mary Frances (2003) "Toward a Globally Harmonized System – Negotiating to Promote Public Health, Environmental Protection, and International Trade", *Fletcher Forum of World Affairs Journal*, Vol. 27, pp. 195–206.

Lübbe-Wolf, Gertrude (1991) "Verfassungsrechtliche Fragen der Normsetzung und Normkonkretisierung im Umweltrecht", *Zeitschrift für Gesetzgebung*, Vol. 6, pp. 219–48.

Luhmann, Niklas (1983) *Rechtssoziologie*, 2nd edn (Opladen: Westdeutscher Verlag).

Luhmann, Niklas (1997) *Legitimation durch Verfahren*, 4th edn (Frankfurt/Main: Suhrkamp).

Mahlmann, Wilfried (2000) *Chemikalienrecht – Einführung und Erläuterung* (Stuttgart: Wissenschaftliche Verlagsgesellschaft).

Majone, Giandomenico (1982) "The Uncertainty Logic of Standard Setting", *Zeitschrift für Umweltpolitik & Umweltrecht*, Vol. 5, pp. 305–23.

Majone, Giandomenico (1984) "Science and Trans-Science in Standard Setting", *Science, Technology & Human Values*, Vol. 9, pp. 15–22.

Majone, Giandomenico (1998) "Europe's 'Democratic Deficit' – The Question of Standards", *European Law Journal*, Vol. 4, pp. 5–28.

Marcussen, Martin (2004) "OECD Governance through Soft Law" in Mörth, Ulrika (ed.) *Soft Law in Governance and Regulation – An Interdisciplinary Analysis* (Cheltenham and Northampton: Edward Elgar), pp. 103–28.

Marschall, Stefan (2002) "'Niedergang' und 'Aufstieg' des Parlamentarismus im Zeitalter der Denationalisierung", *Zeitschrift für Parlamentsfragen*, Vol. 33, pp. 377–90.

Martinez, Magdalena M. Martin (1996) *National Sovereignty and International Organizations* (The Hague: Kluwer Law International).

Mayer, Otto (1924) *Deutsches Verwaltungsrecht*, Vol. 1, 3rd edn (Leipzig: Duncker & Humblot).

Mayntz, Renate (1985) *Soziologie der Verwaltung*, 3rd edn (Heidelberg: Müller, Juristischer Verlag).

Mekouar, Mohamed Ali (2000) "Pesticides and Chemicals – The Requirement of Prior Informed Consent" in Shelton, Dinah (ed.) *Commitment and Compliance: The Role of Non-Binding Norms in the International Legal System* (Oxford: Oxford University Press), pp. 146–63.

Mengel, Constanze (2004) *Sozialrechtliche Rezeption Ärztlicher Leitlinien* (Baden-Baden: Nomos).

Mercier, Michel (1981) "Present and Planned Activities of the International Programme on Chemical Safety (IPCS) on Existing Chemicals" in Umweltbundesamt (ed.) *Proceedings of the Workshop on the Control of Existing Chemicals under the Patronage of the Organisation for Economic Co-Operation and Development* (Berlin: Umweltbundesamt), pp. 39–44.

Merkel, Angela (1997) "Chemikaliensicherheit im Rahmen der Agenda 21", *Umwelt*, pp. 133–4.

Merry, Sally Engle (1988) "Legal Pluralism", *Law & Society Review*, Vol. 22, pp. 868–96.

Messner, Dirk (2002) "Nationalstaaten in der Global Governance-Architektur – Wie kann das deutsche politische System Global Governance-tauglich werden?" *INEF Report*, No. 66.

Meuser, Michael and Nagel, Ulrike (1991) "ExperInneninterviews – vielfach erprobt, wenig bedacht – Ein Beitrag zur qualitativen Methodendiskussion" in Garz, Detlef and Kraimer, Klaus (eds) *Qualitativ-empirische Sozialforschung: Konzepte, Methoden, Analysen* (Opladen: Westdeutscher Verlag), pp. 441–71.

Meuser, Michael and Nagel, Ulrike (2006) "Das ExpertInneninterview – Wissenssoziologische Voraussetzungen und methodische Durchführung" in Friebertshäuser, Barbara and Prengel, Annedore (eds) *Forschungsmethoden in der Erziehungswissenschaft* (Weinheim and München: Juventa Verlag), pp. 481–91.

Ministry for the Environment, Nature Conservation and Nuclear Safety of the Federal Republic of Germany (ed.) (1998) *Environmental Code (Umweltgesetzbuch – UGB) – Draft* (Berlin: Duncker & Humblot).

Molineaux, Charles (1997) "Moving Towards a Construction Lex Mercatoria – A Lex Constructionis", *Journal of International Arbitration*, Vol. 14, pp. 55–66.

Möllers, Christoph (2005) "Transnationale Behördenkooperation – Verfassungs- und völkerrechtliche Probleme transnationaler administrativer Standardsetzung", *Zeitschrift für ausländisches öffentliches Recht und Völkerrecht*, Vol. 65, pp. 351–89.

Morgenthau, Hans J. (1973) *Politics Among Nations – The Struggle for Power and Peace*, 5th edn (New York: Alfred A. Knopf).

Mosler, Hermann (1992) "Die Übertragung von Hoheitsgewalt" in Isensee, Josef and Kirchhof, Paul (eds) *Handbuch des Staatsrechts der Bundesrepublik Deutschland*, Vol. VII (Heidelberg: C. F. Müller), pp. 599–646.

Muchlinski, Peter T. (1997) "'Global Bukowina' Examined – Viewing the Multinational Enterprise as a Transnational Law-Making Community" in Teubner, Gunther (ed.) *Global Law Without a State* (Aldershot: Dartmouth), pp. 79–108.

Münzing, Ekkehard and Pilz, Volker (1998) "Der Auswärtige Ausschuß des Deutschen Bundestage – Aufgaben, Organisation und Arbeitsweise" *Zeitschrift für Parlamentsfragen*, Vol. 29, pp. 575–604.

Nanda, Ved P. and Bailey, Bruce C. (1989) "Nature and Scope of the Problem" in Handl, Günther and Lutz, Robert E. (eds) *Transferring Hazardous Technologies and Substances: The International Legal Challenge* (London: Graham & Trotman), pp. 3–39.

National Institute for Minamata Disease (2008), http://www.nimd.go.jp/english/index.html, date accessed 16 November 2008.

Neuhold, Hanspeter (2005) "The Inadequacy of Law-Making by International Treaties – 'Soft Law' as an Alternative?" in Wolfrum, Rüdiger and Röben, Wolfgang (eds) *Developments of International Law in Treaty Making* (Berlin: Springer, 2005), pp. 39–52.

Nollkämper, André (2006) "Responsibility of Transnational Corporations in International Environmental Law: Three Perspectives" in Winter, Gerd (ed.) *Multilevel Governance of Global Environmental Change – Perspectives from Science, Sociology and the Law* (Cambridge: Cambridge University Press), pp. 178–99.

Nye, Joseph S. (2004) *Soft Power – The Means to Success in World Politics* (New York: PublicAffairs).

O'Connell, Mary Ellen (1993) "Enforcing the New International Law of the Environment", *German Yearbook of International Law*, Vol. 35, pp. 293–332.

O'Connell, Mary Ellen (2000) "The Role of Soft Law in a Global Order" in Shelton, Dinah (ed.) *Commitment and Compliance – The Role of Non-Binding Norms in the International Legal System* (Oxford: Oxford University Press), pp. 100–14.

Obadia, Isaac (2003) "ILO Activities in the Area of Chemical Safety", *Toxicology*, Vol. 190, pp. 105–15.

Odendahl, Kerstin (1998) *Die Umweltpflichtigkeit der Souveränität – Reichweite und Schranken territorialer Souveränitätsrechte über die Umwelt und die Notwendigkeit eine veränderten Verständnisses staatlicher Souveränität* (Berlin: Duncker & Humblot).

OECD (1998) *Honeybees, Acute Oral Toxicity Test (Original Guideline, adopted 21st September 1998)* (Paris: OECD).

OECD (2000) *The OECD Guidelines for Multinational Enterprises: Revision 2000*, http://www.olis.oecd.org/olis/2000doc.nsf/LinkTo/NT00002916/$FILE/00082259.PDF, date accessed 16 November 2008.

OECD (2001) *OECD Environmental Outlook for the Chemicals Industry*, http://www.oecd.org/dataoecd/7/45/2375538.pdf, date accessed 16 November 2008.

OECD (2004a) *Description of OECD Work on Investigation of High Production Volume Chemicals* (October 2004), http://www.oecd.org/document/21/0,3343,en_264 9_34379_1939669_1_1_1_1,00.html, date accessed 16 November 2008.

OECD (2004b) *The 2004 OECD List of High Production Volume Chemicals*, http://www.oecd.org/dataoecd/55/38/33883530.pdf, date accessed 16 November 2008.

OECD (2005) *OECD Environment Programme*, http://www.oecd.org/data oecd/11/45/38135367.pdf, date accessed 16 November 2008

OECD (2006) *Series on the Test Guidelines Programme No. 1: Guidance Document for the Development of OECD Guidelines for Testing of Chemicals*, http://www.oecd.org/dataoecd/2/24/40472761.pdf, date accessed 16 November 2008.

OECD (2007) *OECD Guidelines for Testing of Chemicals: Full List of Test Guidelines* (August 2007), http://www.oecd.org/dataoecd/9/11/33663321.pdf, date accessed 16 November 2008.

OECD (2008a) *Manual for Investigation of HPV Chemicals*, http://www.oecd.org/document/7/0,2340,en_2649_34379_1947463_1_1_1_1,00.html, date accessed 16 November 2008.

OECD (2008b) *OECD's Environment, Health and Safety Programme*, http://www.oecd.org/dataoecd/18/0/1900785.pdf, date accessed 16 November 2008.

Oeter, Stefan (2002) "Souveränität – ein überholtes Konzept?" in Cremer, Hans-Joachim, Giegerich, Thomas and Richter, Dagmar (eds) *Tradition und Weltoffenheit des Rechts: Festschrift für Helmut Stein* (Berlin: Springer), pp. 259–90.

Ohlhoff, Stefan (1999) "Beteiligung von Verbänden und Unternehmen im WTO-Streitbeilegungsverfahren – Das Shrimp-Turtle-Verfahren als Wendepunkt?", *Europäische Zeitschrift für Wirtschaftsrecht*, Vol. 10, pp. 139–44.

Olsen, Marco A. (2003) *Analysis of the Stockholm Convention on Persistent Organic Pollutants* (Dobbs Ferry: Oceana Publications).

Oppermann, Thomas (2005) *Europarecht*, 3rd edn, rev. (München: C. H. Beck).

Ossenbühl, Fritz (1993) "Maastricht und das Grundgesetz – Eine verfassungsrechtliche Wende?", *Deutsches Verwaltungsblatt*, Vol. 108, pp. 629–37.

Ott, Hermann (1998) *Umweltregime im Völkerrecht – Eine Untersuchung zu neuen Formen internationaler institutionalisierter Kooperation am Beispiel der Verträge zum Schutz der Ozonschicht und grenzüberschreitender Abfallverbringung* (Baden-Baden: Nomos).

Pallemaerts, Marc (2003) *Toxics and Transnational Law – International and European Regulation of Toxic Substances As Legal Symbolism* (Oxford: Hart).

Perrez, Franz-Xaver (2000) *Cooperative Sovereignty – From Independence to Interdependence in the Structure of International Environmental Law* (The Hague: Kluwer Law International).

Peters, Anne (2001) *Elemente einer Theorie der Verfassung Europas* (Berlin: Duncker & Humblot).

Pfeil, Norbert, Gerner, Ingrid and Vormann, Kirsten (2000) "Stand und Auswirkungen der globalen Harmonisierung der Einstufung und Kennzeichnung gefährlicher Stoffe und Güter", *Chemie Ingenieur Technik*, Vol. 72, pp. 305–12.

Picciotto, Sol (1996) "Networks in International Economic Integration – Fragemented States and the Dilemmas of Neo-Liberalism", *Northwestern Journal of International Law and Business*, Vol. 17, pp 1014–56.

Picciotto, Sol (1997) "Fragmented States and International Rule of Law", *Social and Legal Studies*, Vol. 6, pp. 259–79.

Pospíšil, Leopol (1982) *Anthropologie des Rechts – Recht und Gesellschaft in archaischen und modernen Kulturen* (München: C. H. Beck).

Pratt, Iona S. (2002) "Global Harmonisations of Classification and Labelling of Hazardous Chemicals", *Toxicology Letters*, Vol. 128, pp. 5–15.

Quaritsch, Helmut (1987) "Legalität, Legitimität" in Herzog, Roman, Kunst, Hermann, Schlaich, Klaus and Schneemelcher, Wilhelm (eds), *Evangelisches Staatslexikon*, Vol. I, 3rd edn, rev. (Stuttgart: Kreuz Verlag), pp. 1989–96.

Rau-Mentzen, Blanca L. and von Koppenfels, Georg (1995) "UNIDO – United Nations Industrial Development Organization" in Wolfrum, Rüdiger (ed.) *United Nations – Law, Policies and Practice* (München and Dordrecht: C. H. Beck; Nijhoff), pp. 1329–34.

Raustiala, Kal (2003) "The Architecture of International Cooperation – Transgovernmental Networks and the Future of International Law", *Virginia Journal of International Law*, Vol. 43, pp. 1–92.

Rehbinder, Eckhard (2003) "Allgemeine Regelungen – Chemikalienrecht" in Rengeling, Hans-Werner (ed.) *Handbuch zum europäischen und deutschen Umweltrecht*, Vol. II/1, 2nd edn (Köln: Heymanns), pp. 547–625.

Reinhard, Wolfgang (2000) *Geschichte der Staatsgewalt – Eine vergleichende Verfassungsgeschichte Europas von den Anfangen bis zur Gegenwart*, 2nd edn (München: C. H. Beck).

Reinicke, Wolfgang H. and Witte, Jan Martin (2000) "Interdependence, Globalization, and Sovereignty – The Role of Non-binding International Legal Accords" in Shelton, Dinah (ed.) *Commitment and Compliance: The Role of Non-Binding Norms in the International Legal System* (Oxford: Oxford University Press), pp. 75–100.

Riedel, Eibe (1991) "Standards and Sources – Farewell to the Exclusivity of the Sources Triad in International Law", *European Journal of International Law*, Vol. 2, pp. 58–84.

Rinken, Alfred (1996) "Demokratie und Hierarchie – Zum Demokratieverständnis des Zweiten Senats des Bundesverfassungsgerichts", *Kritische Vierteljahresschrift für Gesetzgebung und Rechtswissenschaft*, Vol. 79, pp. 282–309.

Rippen, Gerd (1987) *Handbuch Umweltchemikalien Stoffdaten, Prüfverfahren, Vorschriften* (Landsberg/Lech: Ökomed).

Risikokommission (ad hoc-Kommission "Neuordnung der Verfahren und Strukturen zur Risikobewertung und Standardsetzung im gesundheitlichen

254   *Bibliography*

Umweltschutz der Bundesrepublik Deutschland") (2003) *Abschlussbericht der Risikokommission* (Salzgitter: Bundesamt für Strahlenschutz).

Rittberger, Volker (1995) "UNITAR – United Nations Institute for Training and Research" in Wolfrum, Rüdiger (ed.) *United Nations – Law, Policies and Practice* (München and Dordrecht: C. H. Beck;Nijhoff), pp. 1335–40.

Robé, Jean-Phillipe (1997) "Multinational Enterprises – The Constitution of a Pluralistic Legal Order" in Teubner, Gunther (ed.) *Global Law Without the State* (Aldershot: Dartmouth), pp. 45–77.

Röhl, Klaus F. (1987) *Rechtssoziologie* (Köln: Heymann). Röhl, Klaus F. and Magen, Stefan (1996) "Die Rolle des Rechts im Prozeß der Globalisierung", *Zeitschrift für Rechtssoziologie*, Vol. 17, pp. 1–57.

Roller, Gerhard (2003) "Komitologie und Demokratie – Die Brüsseler Durchführungsausschüsse auf neuer Grundlage", *Kritische Vierteljahresschrift für Gesetzgebung und Rechtswissenschaft*, Vol. 86, pp. 249–78.

Rose-Ackerman, Susan (1994) "American Administrative Law under Siege", *Harvard Law Review*, Vol. 107, pp. 1279–302.

Rosenau, James N. (1992) "Governance, Order, and Change in World Politics" in Rosenau, James N. and Czempiel, Ernst-Otto (eds) *Governance without Government: Order and Change in World Politics* (Cambridge: Cambridge University Press), pp. 1–29.

Rosenstock, Robert (1971) "The Declaration of Principles of International Law Concerning Friendly Relations – A Survey", *American Journal of International Law*, Vol. 65, pp. 713–35.

Ross, Jenifer (1999) "Legally Binding Prior Informed Consent", *Colorado Journal of International Environmental Law and Policy*, Vol. 10, pp. 499–529.

Rousseau, Jean-Jacques (2004) *Vom Gesellschaftsvertrag oder Grundsätze des Staatsrechts* (Stuttgart: Reclam).

Sachverständigenrat für Umweltfragen (1996) *Umweltgutachten 1996 – Zur Umsetzung einer dauerhaft umweltgerechten Entwicklung* (Stuttgart: Metzler-Poeschel).

Safe, Stephen (2004) "Endocrine Disruptors and Human Health – Is There a Problem?", *Toxicology*, Vol. 205, pp. 3–10.

Sand, Peter H. (1992) "International Cooperation – The Environmental Experience" in Mathews, Jessica Tuchman (ed.) *Preserving the Global Environment – The Challenge of Shared Leadership* (New York and London: W. W. Norton & Company), pp. 236–79.

Sassen, Saskia (1996) *Losing Control? – Sovereignty in an Age of Globalization* (New York: Columbia University Press).

Scharpf, Fritz W. (1972) *Demokratietheorie zwischen Utopie und Anpassung* (Konstanz: Universitätsverlag).

Scharpf, Fritz W. (1991) "Die Handlungsfähigkeit des Staates am Ende des zwanzigsten Jahrhunderts", *Politische Vierteljahresschrift*, Vol. 32, pp. 621–34.

Scharpf, Fritz W. (1993) "Legitimationsprobleme der Globalisierung – Regieren in Verhandlungssystemen" in Böhret, Carl and Wewer, Göttrik (eds) *Regieren im 21. Jahrhundert* (Opladen: Leske + Budrich), pp. 165–85.

Scharpf, Fritz W. (1999) *Regieren in Europa – Effektiv und Demokratisch?* (Frankfurt/Main: Campus).

Schelsky, Helmut (1961) *Der Mensch in der wissenschaftlichen Zivilisation* (Köln and Opladen: Westdeutscher Verlag).

Schepel, Harm (2005) *The Constitution of Private Governance – Product Standards in the Regulation of Integrating Markets* (Oxford: Hart).
Scheringer, Martin (1999) *Persistenz und Reichweite von Umweltchemikalien* (Weinheim: Wiley-VCH).
Schliesky, Utz (2004) *Souveränität und Legitimation von Herrschaftsgewalt – Die Weiterentwicklung von Begriffen der Staatslehre und des Staatsrechts im europäischen Mehrebenensystem* (Tübingen: Mohr Siebeck).
Schmidt, Walter (1971) "Programmierung von Verwaltungsentschiedungen", *Archiv des öffentlichen Rechts*, Vol. 96, pp. 321–54.
Schmidt-Aßmann, Eberhard (1991) "Verwaltungslegitimation als Rechtsbegriff", *Archiv des öffentlichen Rechts*, Vol. 116, pp. 330–90.
Schmidt-Aßmann, Eberhard (2001) "Government According to Law" in König, Klaus and Siedentopf, Heinrich (eds) *Public Administration in Germany* (Baden-Baden: Nomos), pp. 305–20.
Schmitt Glaeser, Walter (1973) "Partizipation an Verwaltungsentscheidungen", *Veröffentlichungen der Vereinigung der Deutschen Staatsrechtslehrer*, Vol. 31, pp. 179–265.
Schneider, Volker (1988) *Politiknetzwerke der Chemikalienkontrolle – Eine Analyse der transnationalen Politikentwicklung* (Berlin and New York: de Gruyter).
Schneider, Volker (1989) "Transnationale Chemikalienkontrolle – Internationale Technikentwicklung in einer Kontroll-Lücke?" in Albrecht, Ulrich (ed.) *Technikkontrolle und internationale Politik: Die internationale Steuerung von Technologietransfers und ihre Folgen* (Opladen: Westdeutscher Verlag), pp. 195–219.
Scholte, Jan Aart (2002) "Civil Society and Democracy in Global Governance", *Global Governance*, Vol. 8, pp. 281–304.
Schreuer, Christoph (1983) "Die innerstaatliche Anwendung von international-alem 'soft law' aus rechtsvergleichender Sicht", *Österreichische Zeitschrift für öffentliches Recht und Völkerrecht*, Vol. 34, pp. 243–60.
Schreuer, Christoph (1993) "The Waning of the Sovereign State – Towards a New Paradigm for International Law?", *European Journal of International Law*, Vol. 4, pp. 447–75.
Schrijver, Nico (1990) "The Changing Nature of State Sovereignty", *British Yearbook of International Law*, Vol. 70, pp. 65–98.
Schulte, Martin (1998) "Verfassungsrechtliche Beurteilung der Umweltnormung" in Rengeling, Hans-Werner (ed.) *Umweltnormung – Deutsche, europäische und internationale Rechtsfragen* (Köln: Heymanns), pp. 165–86.
Schuppert, Gunnar Folke (1977) "Bürgerinitiativen als Bürgerbeteiligung an staatlichen Entscheidungen – Verfassungstheoretische Aspekte politischer Beteiligung", *Archiv des öffentlichen Rechts*, Vol. 102, pp. 369–409.
Schuppert, Stefan (1998) *Neue Steuerungselemente im Umweltvölkerrecht am Beispiel des Montrealer Protokolls und des Klimaschutzrahmenzübereinkommens: Kosteneffektivität und Innovationswirkungen als Grundsätze in internationalen Verträgen* (Berlin: Springer).
Schütz, Hans-Joachim (1995) "FAO – Food and Agriculture Organization" in Wolfrum, Rüdiger (ed.) *United Nations – Law, Policies and Practice* (München and Dordrecht: C. H. Beck; Nijhoff), pp. 499–522.
Seidl-Hohenveldern, Ignaz (1995) "Internationale Organisationen aufgrund von soft law" in Beyerlin, Ulrich (ed.) *Recht zwischen Umbruch und*

Bewahrung: Völkerrecht, Europarecht und Staatsrecht: Festschrift für Rudolf Bernhardt (Berlin: Springer), pp. 229–39.

Seidl-Hohenveldern, Ignaz and Loibl, Gerhard (2000) *Das Recht der Internationalen Organisationen einschließlich der supranationalen Gemeinschaften* (Köln: Heymanns).

Silk, Jennifer C. (2003) "Development of a Globally Harmonized System for Hazard Communication", *International Journal of Hygiene and Environmental Health*, Vol. 206, pp. 447–52.

Singh, Mahendra P. (2001) *German Administrative Law in Common Law Perspective*, 2nd edn (Berlin: Springer).

Slaughter, Anne-Marie (1997) "The Real New World Order", *Foreign Affairs*, Vol. 76, pp. 183–97.

Slaughter, Anne-Marie (2000a) "Agencies on the Loose? – Holding Government Networks Accountable" in Berman, George A., Herdegen, Matthias and Lindseth, Peter L. (eds) *Transatlantic Regulatory Cooperation: Legal Problems and Political Prospects* (Oxford: Oxford University Press), pp. 521–46.

Slaughter, Anne-Marie (2000b) "Government Networks – The Heart of the Liberal Democratic Order" in Fox, Gregory H. and Roth, Brad R. (eds) *Democratic Governance and International Law* (Cambridge: Cambridge University Press), pp. 199–235.

Slaughter, Anne-Marie (2004) *A New World Order* (Princeton, NJ: Princeton University Press).

Snyder, Francis (2000) "Global Economic Networks and Global Legal Pluralism" in Berman, George A., Herdegen, Matthias and Lindseth, Peter L. (eds) *Transatlantic Regulatory Cooperation: Legal Problems and Political Prospects* (Oxford: Oxford University Press), pp. 99–115.

Sommermann, Karl-Peter (1997) *Staatsziele und Staatszielbestimmungen* (Tübingen: Mohr Siebeck).

Sousa Santos, Boaventura de (2002) *Toward a New Legal Common Sense – Law, Globalization, and Emancipation*, 2nd edn (London: Butterworths).

Spieker gen. Döhmann, Indra (2003) "US-amerikanisches Chemikalienrecht im Vergleich" in Rengeling, Hans-Werner (ed.) *Umgestaltung des deutschen Chemikalienrechts durch europäische Chemikalienpolitik* (Köln: Heymanns), pp. 151–98.

Spielmann, Horst (2004) "Internationale Regulation toxikologischer Prüfmethoden" in Reichl, Franz-Xaver and Schwenk, Michael (eds) *Regulatorische Toxikologie* (Berlin: Springer), pp. 140–7.

Sternberger, Dolf (1967) "Max Webers Lehre von der Legitimität – Eine kritische Betrachtung" in Röhrich, Wilfried (ed.) *Macht und Ohnmacht des Politischen: Festschrift zum 65. Geburtstag von Michael Freund* (Köln and Berlin: Kiepenheuer & Wietsch), pp. 111–26.

Stewart, Richard B. (2005) "US Administrative Law – A Model for Global Administrative Law?", *Law and Contemporary Problems*, Vol. 68, pp. 63–108.

Stockholm Convention on Persistent Organic Pollutants (2008a) *Guidance*, http://www.pops.int/documents/guidance, date accessed 16 November 2008.

Stockholm Convention on Persistent Organic Pollutants (2008b) *Chemicals In Review Process*, http://www.pops.int/documents/meetings/poprc/chem_review.htm, date accessed 16 November 2008.

Stoll, Peter-Tobias (1996) "The International Environmental Law of Cooperation" in Wolfrum, Rüdiger (ed.) *Enforcing Environmental Standards: Economic Mechanisms as Viable Means?* (Berlin: Springer), pp. 39–93.

Strand, Jakob and Asmund, Gert (2003) "Tributyltin Accumulation and Effects in Marine Molluscs from West Greenland", *Environmental Pollution*, Vol. 123, pp. 31–7.

Suy, Eric (2002) "New Players in International Relations" in Kreijen, Gerard (ed.) *State, Sovereignty, and International Governance* (Oxford: Oxford University Press), pp. 373–86.

Teubner, Gunther (1997) "'Global Bukowina' – Legal Pluralism in the World Society" in Teubner, Gunther (ed.) *Global Law Without the State* (Aldershot: Dartmouth), pp. 3–28.

Teubner, Gunther (2000) "Privatregimes – Neo-Spontanes Recht und duale Sozialverfassungen in der Weltgesellschaft?" in Simon, Dieter and Weiss, Manfred (eds) *Zur Autonomie des Individuums: Liber Amicorum Spiros Simitis* (Baden-Baden: Nomos), pp. 437–53.

Thürer, Daniel (1999) "The Emergence of Non-Governmental Organizations and Transnational Enterprises in International Law and the Changing Role of the State" in Hofmann, Rainer (ed.) *Non-State Actors as New Subjects of International Law: International Law – From the Traditional State Order Towards the Law of the Global Community, Proceedings of an International Symposium of the Kiel Walther-Schücking-Institute of International Law March 25 to 28, 1998, Berlin* (Berlin: Duncker & Humblot), pp. 37–58.

Thürer, Daniel (2000) "Soft Law" in Bernhardt, Rudolf (ed.) *Encyclopedia of Public International Law*, Vol. 4 (Amsterdam: North Holland/ Elsevier), pp. 452–60.

Tietje, Christian (1999) "The Changing Legal Structure of International Treaties as an Aspect of an Emerging Global Governance Architecture", *German Yearbook of International Law*, Vol. 41, pp. 26–55.

Tietje, Christian (2001) *Internationalisiertes Verwaltungshandeln* (Berlin: Duncker & Humblot).

Tietje, Christian (2002a) "The Duty to Cooperate in International Economic Law and Related Areas" in Delbrück, Jost (ed.) *International Law of Cooperation and State Sovereignty: Proceedings of an International Symposium of the Kiel Walther-Schücking-Institute of International Law, May 23–26, 2001,* (Berlin: Duncker & Humblot), pp. 45–65.

Tietje, Christian (2002b) "Transnationales Wirtschaftsrecht aus öffentlich-rechtlicher Perspektive", *Zeitschrift für Vergleichende Rechtswissenschaft*, Vol. 101, pp. 404–20.

Tietje, Christian (2003a) "Das Übereinkommen über technische Handelshemmnisse" in Prieß, Hans-Joachim and Berrisch, Georg M. (eds) *WTO-Handbuch* (München: C. H. Beck), pp. 273–325.

Tietje, Christian (2003b) "Recht ohne Rechtsquellen?", *Zeitschrift für Rechtssoziologie*, Vol. 24, pp. 27–42.

Töller, Annette Elisabeth (2002) *Komitologie – Theoretische Bedeutung und praktische Funktionsweise von Durchführungsausschüssen der Europäischen Union am Beispiel der Umweltpolitik* (Opladen: Leske + Budrich).

Tomuschat, Christian (1972) *Verfassungsgewohnheitsrecht? – Eine Untersuchung zum Staatsrecht der Bundesrepublik Deutschland* (Heidelberg: Carl Winter).

Tomuschat, Christian (1976) "Die Charta der wirtschaftlichen Rechte und Pflichten – Zur Gestaltungskraft von Deklarationen der UN Generalversammlung", *Zeitschrift für ausländisches öffentliches Recht und Völkerrecht*, Vol. 36, pp. 444–91.

Tomuschat, Christian (1999) "International Law: Ensuring the Survival of Mankind on the Eve of a New Century – General Course on Public International Law", *Recueil des Cours*, Vol. 281, pp. 9–438.

Triepel, Heinrich (1899) *Völkerrecht und Landesrecht* (Leipzig: C. L. Hirschfeld).

UNECE (2008) *GHS: Status of implementation*, http://www.unece.org/trans/danger/publi/ghs/implementation_e.html, date accessed 16 November 2008.

UNEP Chemicals (2001) *International Activities Related to Chemicals – Overview of International Agreements/Instruments, Organisations and Programmes Concerning Chemicals Management*, 3rd edn (Geneva: UNEP Chemicals).

UNEP (2008) *OECD Initial Assessment Reports For High Production Volume Chemicals including Screening Information Datasets (SIDS)*, http://www.chem.unep.ch/irptc/sids/oecdsids/sidspub.html, date accessed at 16 November 2008.

UNIDO (2008a) *Capacity Building in the Area of POPs Elimination*, http://www.unido.org/doc/29418, date accessed 16 November 2008.

UNIDO (2008b) *UNIDO and Chemicals*, http://www.unido.org/doc/29685, date accessed 16 November 2008.

UNITAR (2008) *UNITAR's Chemicals and Waste Management Programme*, http://www2.unitar.org/cwm/cwmoverview.html, date accessed 16 November 2008.

United States Department of Labor (2008) *Harmonization of Hazard classification and Labeling – Questions and Answers*, http://www.osha.gov/SLTC/hazard-communications/global_questions_answers.html, date accessed 16 November 2008.

United States Department of State (2007) *Treaties in Force – A List of Treaties and Other International Agreements of the United States in Force on January 1, 2007, Section 1 – Bilateral Treaties*, http://www.state.gov/documents/organization/83046.pdf, date accessed 16 November 2008.

Van der Lugt, Cornis (2002) "Thirty Years After Stockholm – What Role for State Sovereignty" in Biermann, Frank, Brohm, Rainer and Dingwerth, Klaus (eds) *Proceedings of the 2001 Berlin Conference on the Human Dimensions of Global Environmental Change "Global Environmental Change and the Nation State"* (Potsdam: Potsdam Institute for Climate Research), pp. 226–35.

Van Staden, Alfred and Vollaard, Hans (2002) "The Erosion of State Sovereignty – Towards a Post Territorial World?" in Kreijen, Gerard (ed.) *State, Sovereignty, and International Governance*, (Oxford: Oxford University Press), pp. 165–84.

VanDorn, Heather M. (1998) "The Rotterdam Convention", *Colorado Journal of International Environmental Law and Policy*, Vol. 10, pp. 281–90.

Verdross, Alfred and Simma, Bruno (1984) *Universelles Völkerrecht – Theorie und Praxis*, 3rd edn (Berlin: Duncker & Humblot).

Vogel, Klaus (1964) *Die Verfassungsentscheidung des Grundgesetzes für eine internationale Zusammenarbeit – Ein Diskussionsbeitrag zu einer Frage der Staatstheorie sowie des geltenden Staatsrechts* (Tübingen: J. C. B. Mohr (Paul Siebeck)).

von Bogdandy, Armin (2000) *Gubernative Rechtsetzung – Eine Neubestimmung der Rechtsetzung und des Regierungssystems unter dem Grundgesetz in der Perspektive gemeineuropäischer Dogmatik* (Tübingen: Mohr Siebeck).

Vos, Ellen (1997) "The Rise of Committees", *European Law Journal*, Vol. 3, pp. 210–29.

Wagner, Burkhard O. (1998) "Das neue Chemikalienprogramm von UNEP Chemicals", *Umweltwissenschaften und Schadstoff-Forschung – Zeitschrift für Umweltchemie und Ökotoxikologie*, Vol. 10, pp. 245–53.

Wallace, Cynthia Day (2002) *The Multinational Enterprise and Legal Control – Host State Sovereignty in an Era of Economic Globalization*, 2nd edn (The Hague: Martinus Nijhoff).

Warning, Michael and Winter, Gerd (2003) "Ansätze zu einer globalen Chemikalienregulierung" in Rengeling, Hans-Werner (ed.) *Umgestaltung des deutschen Chemikalienrechts durch europäische Chemikalienpolitik* (Köln: Carl Heymanns Verlag), pp. 241–74.

Weber, Max (1978a) *Economy and Society – An Outline of Interpretative Sociology*, Vol. I (Berkeley: University of California Press).

Weber, Max (1978b) *Economy and Society – An Outline of Interpretative Sociology*, Vol. II (Berkeley: University of California Press).

Weil, Prosper (1983) "Towards Relative Normativity in International Law?", *American Journal of International Law*, Vol. 77, pp. 413–42.

Weiler, Joseph H. H. (1997) "The Reformation of European Constitutionalism", *Journal of Common Market Studies*, Vol. 35, pp. 97–131.

Weiss, Linda (1997) "Globalization and the Myth of the Powerless State", *New Left Review*, pp. 3–27.

Weiss, Linda (1998) *The Myth of the Powerless State – Governing the Economy in a Global Era* (Cambridge and Oxford: Polity Press).

Weiss, Linda (2003a) "Bringing Domestic Institutions Back In" in Weiss, Linda (ed.) *States in the Global Economy: Bringing Domestic Institutions Back In* (Cambridge and New York: Cambridge University Press), pp. 1–33.

Weiss, Linda (2003b) "Is the State Being 'Transformed' by Globalisation?" in Weiss, Linda (ed.) *States in the Global Economy: Bringing Domestic Institutions Back In* (Cambridge and New York: Cambridge University Press), pp. 293–317.

Wessels, Wolfgang (2000) *Die Öffnung des Staates – Modelle und Wirklichkeit grenzüberschreitender Verwaltungspraxis 1960–1995* (Opladen: Leske + Budrich).

Wilson, Woodrow (1984) "An Address to the Joint Session of Congress, January 8, 1918" in Link, Arthur S. (ed.) *The Papers of Woodrow Wilson*, Vol. 45, (Princeton, NJ: Princeton University Press), pp. 534–9.

Winter, Gerd (1986) "Einführung", in Winter, Gerd (ed.) *Grenzwerte – Interdisziplinäre Untersuchungen zu einer Rechtsfigur des Umwelt-, Arbeits- und Lebensmittelschutzes* (Düsseldorf: Werner), pp. 1–25.

Winter, Gerd (1996) "Kompetenzen der Europäischen Gemeinschaft im Verwaltungsvollzug" in Lübbe-Wolff, Gertrude (ed.) *Der Vollzug des europäischen Verwaltungsrechts* (Berlin: Erich Schmidt Verlag), pp. 107–30.

Winter, Gerd (2005) "Kompetenzverteilung in der Europäischen Mehrebenenverwaltung", *Europarecht*, Vol. 40, pp. 255–76.

Winter, Gerd (2007) "Dangerous Chemicals: A Global Problem on Its Way to Global Governance" in Führ, Martin, Wahl, Rainer and von Wilmowsky, Peter (eds) *Umweltrecht und Umweltwissenschaft – Festschrift für Eckard Rehbinder* (Berlin: Erich Schmidt Verlag).

Wissenschaftlicher Beirat der Bundesregierung Globale Umweltveränderungen (2001) *Welt im Wandel – Neue Strukturen Globaler Umweltpolitik* (Berlin: Springer).

Wolf, Rainer (1987) "Zur Antiquiertheit des Rechts in der Risikogesellschaft", *Leviathan*, Vol. 23, pp. 357–91.

Wolfrum, Rüdiger (1990) "Purposes and Principles of International Environmental Law", *German Yearbook of International Law*, Vol. 33, pp. 308–30.

Wolfrum, Rüdiger (1997) "Kontrolle der auswärtigen Gewalt", *Veröffentlichungen der Vereinigung der Deutschen Staatsrechtslehrer*, Vol. 56, pp. 39–66.

Wolgast, Ernst (1923) "Die auswärtige Gewalt des Deutschen Reiches – unter besonderer Berücksichtigung des Auswärtigen Amtes", *Archiv des öffentlichen Rechts*, Vol. 5, pp. 1–112.

World Bank (2002) *Toxics and Poverty – The Impact of Toxic Substances On the Poor in Developing Countries* (Washington, DC: World Bank).

World Bank (2008) *Gross Domestic Product, PPP*, http://siteresources.worldbank.org/DATASTATISTICS/Resources/GDP_PPP.pdf, date accessed 16 November 2008.

WTO (2008) *International Trade Statistics 2008: II. Merchandise Trade By Product*, http://www.wto.org/english/res_e/statis_e/its2008_e/its08_merch_trade_product_e.pdf, date accessed 16 November 2008.

Würtenberger, Thomas (1982) "Legitimität, Legalität" in Brunner, Otto, Conze, Werner and Koselleck, Reinhart (eds) *Geschichtliche Grundbegriffe: historisches Lexikon zur politisch-sozialen Sprache in Deutschland*, Vol. III (Stuttgart: Klett-Cotta), pp. 677–740.

Würtenberger, Thomas (1986) "Legitimationsmuster von Herrschaft im Laufe der Geschichte", *Juristische Schulung*, Vol. 26, pp. 344–9.

WWF (2008a) *A History of a Global Environmental Conservation Organization*, http://www.panda.org/about_wwf/who_we_are/history/index.cfm, date accessed 16 November 2008.

WWF (2008b) *DetoX Campaign: Reveal the Truths about Chemicals*, http://www.panda.org/about_wwf/where_we_work/europe/what_we_do/wwf_europe_environment/initiatives/chemicals/detox_campaign/index.cfm, date accessed 16 November 2008.

WWF (2008c) *How is WWF Run?* http://www.panda.org/about_wwf/who_we_are/offices/organization.cfm, date accessed 16 November 2008.

WWF (2008d) *Towards a Toxcis Free Future*, http://www.panda.org/about_wwf/what_we_do/policy/toxics/index.cfm, date accessed 16 November 2008.

Younes, Maged (2004) "Toxikologische Risikoanalyse", in Reichl, Franz-Xaver and Schwenk, Michael (ed.) *Regulatorische Toxikologie: Gesundheitsschutz, Umweltschutz, Verbraucherschutz* (Berlin: Springer), pp. 46–50.

Zahedi, Nancy S. (1999) "Implementing the Rotterdam Convention – The Challenge of Transforming Aspirational Goals into Effective Controls on Hazardous Pesticides Exports to Developing Countries", *Georgetown International Environmental Law Review*, Vol. 11, pp. 707–38.

Zaring, David (1998) "International Law by Other Means – The Twilight Existence of International Financial Regulatory Organizations", *Texas International Law Journal*, Vol. 33, pp. 281–321.

Zemanek, Karl (1998) "Is the Term 'Soft Law' Convenient?" in Hafner, Georg and Loibl, Gerhard (eds) *Liber Amicorum: Professor Ignaz Seidl-Hohenvelder in Honor of His 80th Birthday* (The Hague: Kluwer Law International), pp. 843–62.

Zippelius, Reinhold (2003) *Allgemeine Staatslehre (Politikwissenschaft)*, 14th edn, rev. (München: C. H. Beck).

Zitkov, Vladimir (2003) "Chlorinated Pesticides – Adlrin, DDT, Endrin, Dieldrin, Mirex" in Fiedler, Heidelore (ed.) *Persistent Organic Pollutants* (Berlin: Springer), pp. 47–90.

Zumbansen, Peer (2000) "Die vergangene Zukunft des Völkerrechts", *Kritische Justiz*, Vol. 34, pp. 46–68.

Zürn, Michael (1998) *Regieren jenseits des Nationalstaats Globalisierung und Denationalisierung als Chance* (Frankfurt/Main: Suhrkamp).

Zürn, Michael (2004) "Global Governance and Legitimacy Problems", *Government and Opposition*, Vol. 39, pp. 260–287.

Cases

Federal Constitutional Court (Bundesverfassungsgericht)

– Judgement of 29 July 1952, 2 BvE 2/51, *Entscheidungen des Bundesverfassungsgerichts (BVerfGE)*, Vol. 1, pp. 372ff.
– Order of 30 April 1963, 2 BvM 1/62, *BVerfGE*, Vol. 16, pp. 27ff.
– Order of 9 June 1971, 2 BvR 255/69, *BVerfGE*, Vol. 31, pp. 145ff.
– Order of 9 May 1972, 1 BvR 518/62 and 308/64, *BVerfGE*, Vol. 33, pp. 125ff.
– Order of 28 October 1975, 2 BvR 883/73 and 379, 497, 526/74, *BVerfGE*, Vol. 40, pp. 237ff.
– Order of 15 February 1978, 2 BvR 134, 268/76, *BVerfGE*, Vol. 47, pp. 253ff.
– Order of 8 August 1978, 2 BvL 8/77, *BVerfGE*, Vol. 49, pp. 89ff.
– Order of 20 December 1979, 1 BvR 385/77, *BVerfGE*, Vol. 53, pp. 30ff.
– Judgement of 25 January 1983, 2 BvE 1, 2, 3, 4/83, *BVerfGE*, Vol. 62, pp. 1ff.
– Order of 22 March 1983, 2 BvR 457/78, *BVerfGE*, Vol. 63, pp. 343ff.
– Judgement of 17 July 1984, 2 BvE 11, 15/83, *BverfGE*, Vol. 67, pp. 100ff.
– Judgement of 17 July 1984, 2 BvE 13/83, *BVerfGE*, Vol. 68, pp. 1ff.
– Order of 22 October 1986, 2 BvR 197/83, *BVerfGE*, Vol. 73, pp. 339ff.
– Order of 8 April 1987, 2 BvR 687/85, *BVerfGE*, Vol. 75, pp. 223ff.
– Judgement of 26 June 1990, 2 BvF 3/89, *BVerfGE*, Vol. 83, pp. 60ff.
– Judgement of 12 October 1993, 2 BvR 2134, 2159/92, *BVerfGE*, Vol. 89, pp. 155ff.
– Order of 24 May 1995, 2 BvF 1/92, *BverfGE*, Vol. 93, pp. 37ff.
– Judgement of 19 and 20 April 1994, 2 BvE 3/92, 5/93, 7/93, 8/93, *BVerfGE*, Vol. 90, pp. 286ff.
– Order of 25 March 1999, 2 BvE 5/99, *BVerfGE*, Vol. 100, pp. 266ff.
– Judgement of 22 November 2001, 2 BvE 6/99, *Neue Juristische Wochenschrift* 2002, Vol. 55, pp. 1559.
– Order of 25 March 2003, 2 BvQ 18/03, *BVerfGE*, Vol. 108, pp. 34ff.

Federal Administrative Court (Bundesverwaltungsgericht)

– Judgement of 19 December 1985, 7 C 65.82, *Entscheidungen des Bundesverwaltungsgerichts (BVerwGE)*, Vol. 72, pp. 300.
– Judgement of 29 April 1988, 7 C 33.87, *BVerwGE*, Vol. 79, pp. 254ff.

– Order of 21 March 1996, 7 B 16495, *Natur und Recht* 1996, Vol. 18, pp. 522.

European Court of Justice

– Judgement of the Court, 5 February 1963, Cs. 26/62, van Gend & Loos, European Court Reports (*ECR*) 1963, pp. 1ff.
– Judgement of the Court, 15 July 1964, Cs. 6/64, Costa v E.N.E.L., European Court Reports (*ECR*) 1964, pp. 585ff.
– Judgement of the Court, 17 December 1970, Cs. 25/70 Einfuhr- und Vorratsstelle v Köster, European Court Reports (*ECR*) 1970, pp. 1161ff.
– Judgement of the Court, 27 September 1988, C 302/87, Parliament v Council, European Court Reports (*ECR*) 1988, pp. 5615ff.
– Judgement of the Court, 27 October 1992, Cs. 240/90, Germany v Commission, European Court Reports (*ECR*)-I 1992, pp. 5383ff.
– Judgement of the Court of First Instance, 19 July 1999, Cs. T-188/97, Rothmans v Commission, European Court Reports (*ECR*)-II 1999, pp. 2463ff.
– Opinion of the Court, 14 December 1991, Opinion 1/91, European Economic Area, European Court Reports (*ECR*)-I 1991, pp. 6079ff.

International Court of Justice

– Judgement of 9 April 1949 – Corfu Channel Case, ICJ Reports 1949, pp. 4ff.
– Advisory Opinion of 13 July 1954 – Effect of awards of compensation made by the UN Administrative Tribunal, ICJ Reports 1954, pp. 47ff.
– Judgement of 5 February 1970 – Barcelona Traction, Light and Power Company, Limited, ICJ Reports 1970, pp. 3ff.
– Judgement of 27 June 1986 – Military and Paramilitary Activities in and against Nicaragua, Nicaragua v. United States of America, ICJ Reports 1986, pp. 14ff.
– Advisory Opinion of 8 July 1996 – Legality of the Threat of Use of Nuclear Weapons, ICJ Reports 1996, pp. 226ff.
– Advisory Opinion of 8 July 1996 – Dissenting Opinion of Judge Weeramantry, Legality of the Threat or Use of Nuclear Weapons, ICJ Reports 1996, pp. 429.

Permanent Court of Arbitration

– Reports of International Arbitral Awards Vol. III, No. LX, Trail Smelter Case (16 April 1938/11 March 1941), pp. 1905ff.

– Report of International Arbitral Awards, Vol. XII, Affaire du Lac Lanoux, (19 November 1956), pp. 281ff.

Permanent Court of International Justice

– Judgement of 7 September 1927, Judgement No. 9 – The Case of the S. S. "Lotus", Permanent Court of International Justice Publications, Series A, Vol. 2, Nos. 9-16 (1927-1928)).

Supreme Court of the United States

– United States Supreme Court, Mistretta v. United States, 488 U.S. 361 (1989).

WTO Dispute Resolution

– Report of the Appellate Body, United States – Import Prohibition of Certain Shrimp and Shrimp Products, 12 October 1998.
– Report of the Appellate Body, European Communities – trade description of sardines, 26 September 2002 (WT/DS231/AB/R).

Legal Acts and Documents

International Primary and Secondary Law

Conventions

- Accord européen relatif au transport international des marchandises dangereuses par voies de navigation intérieures, Geneva, 26 May 2000
- Accord européen relatif au transport international des merchandises dangereuses par route, Geneva, 30 September 1957
- Anti-Fouling Convention: International Convention on the Control of Harmful Anti-fouling Systems on Ships, London, 5 October 2005
- Articles of Agreement of the International Monetary Fund, Washington, 22 July 1944
- CISG: United Nations Convention on Contracts for the International Sale of Goods, Vienna, 11 April 1980
- Convention of Mainz Concerning the Navigation on the Rhine, Mainz, 26 March 1861
- Convention of Mannheim Concerning the Navigation on the Rhine, Mannheim, 17 October 1868
- Convention on the Settlement of Investment Disputes Between States and Nationals of Other States, Washington, 18 March 1965
- COTIF: *Convention relative aux transports internationaux ferroviaires*, Berne, 9 May 1980
- FAO-Convention: Constitution of the Food and Agriculture Organization of the United Nations, Quebec, 16 October 1945
- ICAO-Convention: Convention on International Civil Aviation, Chicago, 7 December 1944
- ILO-Constitution: Constitution of the International Labour Organization, Philadelphia, 10 May 1944
- IMO-Convention: Convention on the Internaitonal Maritime Organization, Geneva, 6 March 1948
- Kyoto Protocol to the United Nations Framework Convention on Climate Change, Kyoto, 11 December 1997
- Montreal Protocol on Substance that Deplete the Ozone Layer, Montreal, 16 September 1987, last amended in Bejing in 1999
- OECD-Convention: Convention on the Organisation for Economic Co-operation and Development, Paris, 14 December 1960

- Ozone Convention: Vienna Convention for the Protection of the Ozone Layer, Vienna, 22 March 1985
- PIC-Convention: Rotterdam Convention on the Prior Informed Consent Procedure for Certain Hazardous Chemicals and Pesticides in International Trade, Rotterdam, 10 September 1998
- POP-Convention: Stockholm Convention on Persistent Organic Pollutants, Stockholm, 22 May 2001
- SOLAS-Convention: Convention for the Safety of Live at Sea, London, 1 November 1974
- SPS-Agreement: Agreement on Sanitary and Phytosanitary Measures, Marrakesh, 14 April 1994
- Strasbourg Convention (revised Convention Concerning the Navigation on the Rhine), Strasbourg, 7 October 1961
- TBT-Agreement: Agreement on Technical Barriers of Trade, Marrakesh, 14 April 1994
- UN-Charter: Charter of the United Nations, San Francisco, 26 June 1945
- UNFCC: United Nations Framework Convention on Climate Change, New York, 9 May 1992
- UNIDO-Constitution: Constitution of the United Nations Development Organization, Vienna, 8 April 1979
- Vienna Final Act (*Acte du Congrès de Vienne*), Vienna, 9 June 1815
- WHO-Constitution: Constitution of the World Health Organization, New York, 22 July 1946
- VCLT: Vienna Convention on the Law of Treaties, Vienna, 22 May 1969

COP – PIC-Convention

- Report of the Conference of the Parties to the Rotterdam Convention on the Prior Informed Consent Procedure for Certain Hazardous Chemicals and Pesticides in International Trade on the work of its first meeting, Geneva, 22 October 2004 (UNEP/FAO/RC/COP.1/33)
- Report of the Conference of the Parties to the Rotterdam Convention on the Prior Informed Consent Procedure for Certain Hazardous Chemicals and Pesticides in International Trade on the work of its third meeting, Geneva, 9–13 October 2006 (UNEP/FAO/RC/COP.3/26)
- Report of the Conference of the Parties to the Rotterdam Convention on the Prior Informed Consent Procedure for Certain Hazardous Chemicals and Pesticides in International Trade on the work of its fourth meeting, Geneva, 27–31 October 2008 (UNEP/FAO/RC/COP.4/24)

– Rules of Procedure for the Conference of the Parties, Rotterdam Convention on the Prior Informed Consent Procedure for Certain Hazardous Chemicals and Pesticides in International Trade Conference of the Parties – First meeting, Geneva, 20–4 September 2004 (UNEP/FAO/RC/COP.1/2)

COP – POP-Convention

– Report of the Conference of the Parties of the Stockholm Convention on Persistent Organic Pollutants on the work of its third meeting, Dakar, 30 April–4 May 2007 (UNEP/POPS/COP.3/30)
– Rules of Procedure for the Conference of the Parties, Conference of the Parties of the Stockholm Convention on Persistent Organic Pollutants – First meeting, Punta del Este, Uruguay, 2–6 May 2005 (UNEP/POPS/COP.1/25)

IMO

– Marine Environment Protection Committee, Resolution on Measures to Control Potential Adverse Impacts Associated with Use of Tributyl Tin Compounds in Anti-Fouling Paints (MEPC 46(30)), 16 November 1990

INC – PIC-Convention

– Report of the Intergovernmental Negotiating Committee for an International Legally Binding Instrument for the Application of the Prior Informed Consent Procedure for Certain Hazardous Chemicals and Pesticides in International Trade, Fourth session, Rome, 20–4 October 1997 (UNEP/FAO/PIC/INC.4/2)

INC – POP-Convention

– Intergovernmental Negotiating Committee for an International Legally Binding Instrument for Implementing International Action on Certain Persistent Organic Pollutants, Intergovernmental Negotiating Committee – First session, Montreal, 29 June–3 July 1998 (UNEP/POPS/INC.1/2)

ILO

– Convention concerning Safety in the use of Chemicals at Work (C170), Geneva, 25 June 1990
– International Labour Conference, Resolution concerning harmonization of systems of classification and labelling for the use of hazardous chemicals at work, Geneva, 1989

- ILO Governing Body (GB.268/LILS/4/1), 268th Session, Entry Into Force of the Revised Memorandum of Understanding Concerning Cooperation in the International Programme on Chemical Safety (UNEP, ILO, WHO), March 1997
- ILO Tripartite Declaration of Principles concerning Multinational Enterprises and Social Policy, Declaration adopted by the Governing Body of the International Labour Office at its 204th Session, Geneva, November 1977
- Lead Poisoning (Women and Children) Recommendation (No. 4), 1919
- White Phosphorus Recommendation (No. 6), 1919
- White Lead (Painting) Convention (C13), 1921

OECD

- Decision of the Council concerning a Special Programme on the Control of Chemicals, C (78) 127(Final), 21 September 1978
- Decision concerning the Mutual Acceptance of Data in the Assessment of Chemicals, C (81) 30/FINAL, 12 May 1981
- Decision-Recommendation of the Council on the Systematic Investigation of Existing Chemicals, C (87) 90/Final, 26 June 1987
- Decision-Recommendation of the Council on Compliance with Principles of Good Laboratory Practice, C (89) 87/Final, 2 October 1989, amended by C (95) 8/Final, 9 March 1995
- Decision-Recommendation of the Council on the Co-operative Investigation and Risk Reduction of Existing Chemicals, C (90) 163, 31 January 1991
- Recommendation of the Council on the Biodegradability of Anionic Synthetic Surface Active Agents, C (71) 83/Final, 13 July 1971
- Recommendation of the Council establishing Guidelines in Respect of Procedure and Requirements for Anticipating the Effects of Chemicals on Man and in the Environment, C (77) 97/Final, 7 July 1977
- Resolution of the Council concerning the Renewal of the Mandate of the Environment Policy Committee [C (2004) 99/REV1], approved by the Council on 9 June 2004 at its 1088th session [C/M (2004) 14, Item 191]

OSCE

- Charter of Paris for a New Europe, Meeting of the Heads of State or Government of the participating States of the Conference on Security and Co-operation in Europe (CSCE), Paris, 19–21 November 1990

UNEP

– Governing Council, Decision 14/27, 17 June 1987
– Governing Council, Decision 18/32 concerning Persistent Organic Pollutants, 25 May 1995
– Governing Council, Decision 19/13C concerning International Action to Protect Human Health and the Environment through Measures which will Reduce and/or Eliminate Emissions and Discharges of Persistent Organic Pollutants, Including the Development of an Internationally Legally Binding Instrument, 7 February 1997
– Governing Council, Decision SS.VII/3, 7th Special Session/Global Ministerial Environment Forum, 15 February 2002

UNITAR

– Statute of the United Nations Institute for Training and Research, as promulgated by the Secretary-General in November 1965 and amended in March 1967, June 1973, June 1979, May 1983, April 1988, December 1989 and December 1999, Geneva 2000

United Nations

General Assembly

– Resolution 1934 (XVIII), 1276th plenary meeting, 11 December 1963
– Resolution 2625 (XXV), 1883rd plenary meeting, 24 October 1970
– Resolution 2997 (XXVII), 2112th plenary meeting, 115 December 1972
– Resolution 3281 (XXIX), 2315th plenary meeting, 12 December 1974
– Resolutions 39/72, 99th plenary meeting, 13 December 1983
– Resolution 40/64 A, 111th plenary meeting, 12 December 1985
– Resolution 41/35 A, 64th plenary meeting, 10 November 1986
– Resolution 45/176 A, 70th plenary meeting, 19 December 1990

ECOSOC

– Resolution 36 (IV), 28 March 1947
– Resolution 468 G (XV), 15 April 1953
– Resolution 645 G (XXIII), 26 April 1957
– Resolution 1996/31, 49th plenary meeting, 25 July 1996
– Resolution 1999/65, 48th plenary meeting, 26 October 1999
– Resolution 2003/64, 49th plenary meeting, 25 July 2003

Secretary-General

– United Nations Recommendations for the Transport of Dangerous Goods/Model Regulations (ST/SG/AC.10/1/Rev.13), 13th revised edition, New York and Geneva 2003, Geneva, 26 April 1957

– UN Secretariat, Note by the Secretariat (ST/SG/AC.10/C.4/2001/7), 27 April 2001
– UN Secretariat, Note by the Secretariat (ST/SG/AC.10/C.4/2001/8), 9 May 2001
– UN Secretariat, Notes by the Secretariat (ST/SG/AC.10/C.4/2001/1), 7 June 2001

WHO

– World Health Assembly, Evaluation of the Effects of Chemicals on Health (WHA 30.47), 30th World Health Assembly, Geneva, 2–19 May 1977
– World Health Assembly, Evaluation of the Effects of Chemicals on Health (WHA 31.28), 31st World Health Assembly, Geneva, 8–24 May 1978
– Executive Board, Evaluation of the Effects of Chemicals on Health (EB.63.R.19), 63rd Session Geneva, 10–26 January 1979

International and Transnational Documents

Stockholm 1972 – United Nations Conference on the Human Environment

– Declaration of the United Nations Conference on the Human Environment, adopted at Stockholm on 16 June 1972

Rio de Janeiro 1992 – United Nations Conference on Environment and Development

– Declaration on Environment and Development, Rio de Janeiro, 3–14 June 1992
– Agenda 21, Rio de Janeiro, 3–14 June 1992

IPCS

– Persistent Organic Pollutants: An Assessment Report on DDT, Aldrin, Dieldrin, Endrin, Chlordane, Heptachlor-Hexachlorobenzene, Mirex, Toxaphene, Polychlorinated biphenyls, Dioxins, and Furans, December 1995

IFCS

– Resolution on the Establishment of an Intergovernmental Forum on Chemical Safety (IPCS/ IFCS/ 94.Res.1), Stockholm, 29 April 1994
– IFCS II, Second Session of the Intergovernmental Forum on Chemical Safety, Final Report (IFCS/FORUM-II/97.25w), Ottawa, 10–14 February 1997

- IFCS III, Third Session of the Intergovernmental Forum on Chemical Safety, Final Report (IFCS/ Forum III/ 23w), Salvador da Bahía, 15–20 October 2000
- IFCS IV, Forum IV, Thought Starter Report to SAICM PrepCom1 (IFCS/ FORUM IV/13w Rev 2), Bangkok, 9–13 November 2003

ICCM

- Strategic Approach to International Chemicals Management Comprising the Dubai Declaration on International Chemicals Management, the Overarching Policy Strategy and the Global Plan of Action, Dubai, 6 February 2006
- Report of the International Conference on Chemicals Management on the Work of Its First Session (SAICM/ICCM.1/7), Dubai, 4–6 February 2006

IOMC

- IOMC MoU: Memorandum of Understanding concerning Establishment of the Inter-Organization Programme for the Sound Management of Chemicals, entered into force 13 March 1995
- IOMC IOCC, Summary Record of the Thirteenth Meeting (IOMC/ IOCC/00.08), Geneva, 22–3 June 2000
- IOMC IOCC, Committee, Summary Record of the Fourteenth Meeting (IOMC/IOCC/01.01 Rev.1), Vienna, 18–19 December 2000
- IOMC IOCC, Summary Record of the Eighteenth Meeting (IOMC/ IOCC/03.44), Geneva, 22–3 January 2002
- CG/HCCS, Revised Terms of Reference and Work Programme (IOMC/ HCS/95), Geneva, 14 January 1996

UNCETDG/GHS – UNSCEGHS – UNSCETDG

- UNCETDG/GHS, Report of the Committee of Experts on Its First Session (ST/SG/AC.10/29), 11–12 December 2002
- UNSCEGHS, Report of the Sub-Committee of Experts on Its Tenth Session (ST/SG/AC.10/C.4/20), 7–8 December 2005
- UNSCEGHS, Implementation of GHS, Implementation through Transport of Dangerous Goods Regulations (UN/SCEGHS/11/INF.2), 14 July 2006

Ad Hoc Joint Working Group on Enhancing Cooperation among the Basel, Rotterdam and Stockholm Convention

- Report of the ad hoc joint working group on enhancing cooperation and coordination among the Basel, Rotterdam and Stockholm

conventions on the work of its first meeting (UNEP/FAO/CHW/RC/
POPS/JWG.1/4), 18 April 2007.

Statutes

European Union/ European Community

– *EC-Treaty*: Treaty establishing the European Community, 24 December
2002 (Official Journal C 325, pp. 1ff.)
– *EU-Treaty*: Treaty on European Union, 24 December 2002 (Official
Journal C 325, pp. 1ff.)
– Council Decision 87/373/EEC of 13 July 1987 laying down the procedures for the exercise of implementing powers conferred on the
Commission (Official Journal L 197, 18 July 1987, pp. 33–5), repealed
by Council Decision 1999/468/EC
– Council Decision 1999/468/EC of 28 June 1999 laying down the
procedures for the exercise of implementing powers conferred on
the Commission (Official Journal L 184, 17 July 1999 pp. 23–6), last
amended by Council Decision 2006/512/EC
– Council Decision 2006/512/EC of 17 July 2006 amending Decision
1999/468/EC laying down the procedures for the exercise of implementing powers conferred on the Commission (Official Journal L
200, 22 July 2006, pp. 11–13)
– Council Directive 67/548/EEC of 27 June 1967 on the approximation
of laws, regulations and administrative provisions relating to the classification, packaging and labelling of dangerous substances (Official
Journal 196, 16 August 1967, pp. 1–98), last amended by Commission
Directive 2004/73/EC of 29 April 2004 adapting to technical progress
for the 29th time Council Directive 67/548/EEC... (Official Journal L
152 , 30 April 2004 pp. 1–316)
– Council Directive 76/769/EEC of 27 July 1976 on the approximation
of the laws, regulations and administrative provisions of the Member
States relating to restrictions on the marketing and use of certain
dangerous substances and preparations (Official Journal L 262, 27
September 1976 pp. 201–3), last amended by Directive 2005/90/EC
of the European Parliament and of the Council of 18 January 2006
amending, for the 29th time, Council Directive 76/769/EEC... (Official
Journal L 33, 4 February 2006, pp. 28–81)
– Council Directive 79/831/EEC of 18 September 1979 amending for the
sixth time Directive 67/548/EEC on the approximation of the laws,
regulations and administrative provisions relating to the classification,
packaging and labelling of dangerous substances (Official Journal L
259, 15 October 1979, pp. 10–28)

- Council Directive 82/501/EEC of 24 June 1982 on the major-accident hazards of certain industrial activities ("Seveso I") (Official Journal L 230, 5 August 1982 pp. 1–18), repealed by Council Directive 96/82/EC of 9 December 1996 on the control of major-accident hazards involving dangerous substances (Official Journal L 10, 14 January 1997, pp. 13–33)
- Commission Directive 93/67/EEC of 20 July 1993 laying down the principles for assessment of risks to man and the environment of substances notified in accordance with Council Directive 67/548/EEC (Official Journal L 227, 8 September 1993, pp. 9–18)
- Council Directive 96/61/EC of 24 September 1996 concerning integrated pollution prevention and control
- Council Directive 96/82/EC of 9 December 1996 on the control of major-accident hazards involving dangerous substances ("Seveso II") (Official Journal L 10, 14 January 1997, pp. 13–33)
- Council Regulation (EEC) No. 793/93 of 23 March 1993 on the evaluation and control of the risks of existing substances (Official Journal L 084, 5 April 1993, pp. 1–75)
- Regulation (EC) No. 304/2003 of the European Parliament and of the Council of 28 January 2003 concerning the export and import of dangerous chemicals (Official Journal L 63, 6 March 2003, pp. 1–26)
- Regulation (EC) No. 782/2003 of the European Parliament and of the Council of 14 April 2003 on the prohibition of organotin compounds on ships (Official Journal L 115, 9 May 2003, pp. 1–11)
- Regulation (EC) No. 850/2004: Regulation (EC) No. 850/2004 of the European Parliament and of the Council of 29 April 2004 on persistent organic pollutants and amending Directive 79/117/EEC (Official Journal L 158, 30 April 2004, pp. 7–49)
- Regulation (EC) No. 1907/2006: Regulation (EC) No. 1907/2006 of the European Parliament and of the Council of 18 December 2006 concerning the Registration, Evaluation, Authorisation and Restriction of Chemicals (REACH), establishing a European Chemicals Agency, amending Directive 1999/45/EC and repealing Council Regulation (EEC) No 793/93 and Commission Regulation (EC) No 1488/94 as well as Council Directive 76/769/EEC and Commission Directives 91/155/EEC, 93/67/EEC, 93/105/EC and 2000/21/EC (Official Journal L 396, 30 December 2006, pp. 1–849)
- Regulation (EC) 440/2008: Council Regulation (EC) No 440/2008 of 30 May 2008 laying down test methods pursuant to Regulation (EC) No 1907/2006 of the European Parliament and of the Council on the Registration, Evaluation, Authorisation and Restriction of Chemicals (REACH) (Official Journal L 142, 31 May 2008, pp. 1–739)

Germany[378]

– *Grundgesetz für die Bundesrepublik Deutschland* (Basic Law for the Federal Republic of Germany), 23 May 1949 (Bundesgesetzblatt 1949, 1), last amended by Art. 1 of the Act of 26 July 2002 (Bundesgesetzblatt I S. 2863)[379]

– *Bundes-Immissionsschutzgesetz – Gesetz zum Schutz vor schädlichen Umwelteinwirkungen durch Luftverunreinigungen, Geräusche, Erschütterungen und ähnliche Vorgänge* (Federal Immission Control Act – Act on the prevention of harmful effects on the environment caused by air pollution, noise, vibration and similar phenomena), as promulgated 26 September 2002 (Bundesgesetzblatt I S. 3830), last amended by Artikel 1 of the Act of 25 June 2005 (Bundesgesetzblatt I S. 1865)[380]

– *Bundesverfassungsgerichtsgesetz – Gesetz über das Bundesverfassungsgericht* (Act concerning the Federal Constitutional Court) as promulgated 11 August 1993 (Bundesgesetzblatt I S. 1473), last amended by Artikel 5 Absatz 2 of the Act of 15 December 2004 (Bundesgesetzblatt I S. 3396)

– *Benzinbleigesetz –* Gesetz zur Verminderung von Luftverunreinigungen durch Bleiverbindungen in Ottokraftstoffen für Kraftfahrzeugmotore (Act on the reduction of air pollution through lead compounds in liquid-fuel spark-ignition engines in automobiles), 5 August 1971 (Bundesgesetzblatt I S. 1234), last amended by Artikel 40 of the Ordinance of 25 November 2003 (Bundesgesetzblatt I S. 2304)

– *Chemikalienverbotsverordnung – Verordnung über Verbote und Beschränkungen des Inverkehrbringens gefährlicher Stoffe, Zubereitungen und Erzeugnisse nach dem Chemikaliengesetz* (Ordinance on the Prohibition and Restriction of the Placing on the Market of Hazardous Substances, Preparations and Articles), as promulgated 13 June 2003 (Bundesgesetzblatt I S. 867, last amended by Artikel 1 of the Act of 11 July 2006 (Bundesgesetzblatt I S. 1575)

– *DDT-Gesetz – Gesetz über den Verkehr mit DDT* (Act on the commerce with DDT) 7 August 1972 (BGBI. I S. 1385), repealed by the *2. Gesetz*

---

[378] An overview of English translations of German Acts and other legal instruments available online is provided by http://www.words-worth.de/robin/german-law. A bibliography of English translations is available at http://www.iuscomp.org/gla/statutes/biblstindex.htm.

[379] English translation at http://www.bundestag.de/interakt/infomat/fremd-sprachiges_material/downloads/ggEn_download.pdf.

[380] English translation at http://www.bmu.bund.de/files/pdfs/allgemein/application/pdf/bimschengl.pdf

zur Änderung des Chemikaliengesetzes (2nd Act amending the Chemicals Act) 25 July 1994 (Bundesgesetzblatt I S. 1703)
- *Gefahrstoffverordnung – Verordnung zum Schutz vor Gefahrstoffen* (Ordinance on Harzardous Substances), 23 December 2004 (Bundesgesetzblatt I S. 3758, 3759), last amended by Artikel 2 of the Ordinance of 11 July 2006 (Bundesgesetzblatt I S. 1575)
- *Gemeinsame Geschäftsordnung der Bundesministerien* (Joint rules of procedure of the federal ministries), resolution of the federal cabinet, 26 July 2000 (GMBl. Nr. 28, S. 525)
- *Gesetz zum Übereinkommen vom 14. Dezember 1960 über die Organisation für Wirtschaftliche Zusammenarbeit und Entwicklung (OECD)* (Act concerning the Convention of 14 December 1960 on the Organisation for Economic Cooperation and Development (OECD)), 16 August 1961 (Bundesgesetzblatt II, S. 1150)
- *Geschäftsordnung der Bundesregierung* (Rules of procedure of the federal government), resolution of the federal cabinet, 11 May 1951 (GMBl. S. 137), last amended 21 November 2002 (GMBl. S. 848)
- *Geschäftsordnung des Deutschen Bundestag* (Rules of Procedure for the German *Bundestag)*, 25 June 1980 (BGBl I 1980, S. 1237), last amended 21 October 2005 (Bundesgesetzblatt S. 3094)[381]
- *Technische Anleitung zum Schutz gegen Lärm – Sechste Allgemeine Verwaltungsvorschrift zum Bundes-Immissionsschutzgesetz* (Technical Instructions on Noise Control – Sixth General Administrative Regulation Pertaining the Federal Immission Control Act), 26 August 1998, (Gemeinsames Ministerialblatt[382] 1998 S. 503)
- *Technische Anleitung zur Reinhaltung der Luft– Erste Allgemeine Verwaltungsvorschrift zum Bundes–Immissionsschutzgesetz* (Technical Instructions on Air Quality Control – First General Administrative Regulation Pertaining the Federal Immission Control Act), 24 July 2002 (Gemeinsames Ministerialblatt 2002, S. 511)
- *Verordnung über elektromagnetische Felder–Sechsundzwanzigste Verordnung zur Durchführung des Bundes-Immissionsschutzgesetzes* [Ordinance on Electromagnetic Fields – 26th Federal Immission Control Ordinance to Implement the Federal Immission Control Act], 16 December 1996 (Bundesgesetzblatt I S. 1966)
- *Verwaltungsgerichtsordnung* (Regulations governing administrative courts), as promulgated 19 March 1991 (Bundesgesetzblatt I S. 686), last amended by Artikel 13 of the Act of 15 July 2006 (Bundesgesetzblatt I S. 1619)

---

[381] English translation at http://www.bundestag.de/htdocs_e/info/rule.pdf.
[382] Joint Ministerial Gazette, issued by the Federal Ministry of the Interior.

Japan

- *Law No. 117*: Law Concerning the Evaluation of Chemical Substances and Regulation of Their Manufacture etc. (Law No. 117), enacted 16 October 1973

New Zealand

- Hazardous Substances and New Organisms Act, Public Act 1996 No. 30, Date of assent 10 June 1996

USA

- Constitution of the United States, adopted 17 September 1787
- Administrative Procedures Act (APA), 5 USC 500 et seq., enacted 11 June 1946
- Toxic Substances Control Act (TSCA), 15 USC 2601 et seq., enacted 11 October 1976
- Code of Federal Regulations, Title 40 – Protection of the Environment
- Memorandum of Understanding regarding U.S. Environmental Protection Agency collaboration in the international program on chemical safety. Signed at Washington and Geneva January 19 and March 19, 1981; entered into force March 19, 1981

Government and Parliamentary Documents

Germany

- Bekanntmachung der Satzung der Ernährungs- und Landwirtschaftsorganisation der Vereinten Nationen [Promulgation of the Constitution of the Food and Agriculture Organization], 13 July 1971 (Bundesgesetzblatt II 1972, S. 1033)
- Bekanntmachung der Satzung der Weltgesundheitsorganisation [Promulgation of the Constitution of the World Health Organization] (Bundesgesetzblatt II 1974, S. 43)
- Bekanntmachung über die Neufassung der Bekanntmachung über die Bildung eines Kerntechnischen Ausschusses [Announcement regarding the revision of the Announcement concerning the Creation of a Nuclear Safety Standards Commission], 20 July 1990 (Bundesanzeiger Nr. 144, 4 August 1990), last amended 22 April 1999 (BundesanzeigerNr. 85, 7 May 1999)
- Deutscher Bundestag, Schlussbericht der Enquete-Kommission – Globalisierung der Weltwirtschaft – Herausforderungen und Antworten, 12 June 2002, Bundestagsdrucksache (*BT-Ds.*) 14/9200

- Deutscher Bundestag, Für eine parlamentarische Mitwirkung im System der Vereinten Nationen, 16 June 2005, Bundestagsdrucksache (*BT-Ds.*) 15/5690
- Deutscher Bundestag, Für eine parlamentarische Dimension im System der Vereinten Nationen, 23 September 2004, Bundestagsdrucksache (*BT-Ds.*) 15/3711,
- Unterrichtung durch die Bundesregierung: Tierschutzbericht 2001 – Bericht über den Stand der Entwicklung des Tierschutzes, 29 March 2001, Bundestagsdrucksache (*BT-Ds.*) 14/5712,
- Unterrichtung durch die Bundesregierung: Tierschutzbericht 2003 – Bericht über den Stand der Entwicklung des Tierschutzes, 23 March 2003, Bundestagsdrucksache (*BT-Ds.*) 15/0723,
- Unterrichtung durch die Bundesregierung: Tierschutzbericht 2005, 27 April 2005, Bundestagsdrucksache (*BT-Ds.*) 15/5405
- Vertrag zwischen der Bundesrepublik Deutschland, vertreten durch den Bundesminister für Wirtschaft, und dem DIN Deutsches Institut für Normung e.V., vertreten durch dessen Präsidenten [Agreement between the Federal Republic of Germany, represented by the Federal Minister for Economic Affairs, and the German Institute of Standardization, represented by its president], 5 July 1975, Beilage zum Bundesanzeiger Nr. 114, 27 June 1975

European Union/ European Community

- Agreement Between the European Parliament and the Commission on procedures for implementing Council Decision 1999/468/EC of 28 June 1999 laying down the procedures for the exercise of implementing powers conferred on the Commission (Official Journal L 256, 10 October 2000, pp. 19–20)
- European Commission, Commission Working Document (SEC (1998) 1986 final), 18 November 1998
- European Commission, List of Committees Which Assist the Commission in the Exercise of Its Implementing Powers (Official Journal C 225, 8 August 2000, pp. 2–18)
- European Commission, White Paper "Strategy on a future Chemicals Policy", (COM (2001) 88/final), 27 February 2001
- European Commission, Report from the Commission on the working of committees during 2004 (COM (2005) 554 final), 10 November 2005
- European Commission, Memo: Q and A on the new Chemicals policy, REACH (MEMO/06/488), Brussel, 13 December 2006

- Proposal for a Regulation of the European Parliament and of the Council on classification, labelling and packaging of substances and mixtures, and amending Directive 67/548/EEC and Regulation (EC) No 1907/2006, COM (2007) 355 final
- European Parliament, Resolution on the Reform of the United Nations, 6 September 2005 (P6_TA(2005)0237)

USA

- Memorandum of Understanding regarding U.S. Environmental Protection Agency collaboration in the international program on chemical safety, signed at Washington and Geneva January 19 and March 19, 1981; entered into force March 19, 1981
- United States House of Representatives Committee on Government Reform – Minority Staff Special Investigations Division, A Special Interest Case Study: The Chemical Industry, the Bush Administration, and European Efforts to Regulate Chemicals, 1 April 2004
- United States General Accounting Office, Testimony Before the Subcommittee on Toxic Substances, Research and Development, Committee on Environment and Public Works, U.S. Senate, Toxic Substances Control Act – Preliminary Observations on Legislative Changes to Make TSCA More Effective (GAO/T-RCED-94-263), 13 July 1994

Interviews

Interview 11 July 2002	Head of division/chemist Ministry
Interview 27 April 2004	Head of department/chemist Government agency
Interview 11 May 2004	Official/toxicologist Government agency
Interview 8 December 2004	Head of unit/chemist Government agency
Interview 15 December 2004	Head of department/toxicologist Government agency
Interview 13 July 2005	Toxicologist Private research institute
Interview (a) 19 July 2005	Official/chemist IO
Interview (b) 19 July 2005	Official/chemist IO
Interview (a) 14 October 2005	Official/chemist IO
Interview (b) 14 October 2005	Official/ political scientist IO
Interview 17 October 2005	Official/ political scientist IO

Index